The powder is shot

Small primer on combination weapons

Horst Decker, Kirchbergstr. 37, 63691 Ranstadt, Germany
ISBN: 979 833 6655834, 228 Fotos
cr.: August 2024
IIX

Weapons have long been collected in a manner similar to jewelry and art, reflecting a combination of functionality and artistic craftsmanship. Historically, weapons were not only tools of war but also status symbols, often elaborately designed and decorated with precious metals, gemstones, and other valuable materials to signify the social standing of their owners. This dual purpose meant that, in addition to their functional roles, these weapons were frequently created with a focus on aesthetic appeal and artistic design.

Moreover, the creation of these weapons required a level of skill and craftsmanship comparable to that of fine jewelry and art, reinforcing their value both as tools of combat and as objects of beauty. The artistic design and decorative elements served to demonstrate the owner's wealth and status, further intertwining the weapon's practical and symbolic purposes

In addition, there were products that attracted attention and interest solely because of their technical design.
'The small primer on combination weapons of the 18th to 20th century ' deals with such weapons.

Although there were already magnificent, courtly combination weapons in the 17th century. The invention of the bayonet initiated a systematic development in which, on the one hand, useful weapon combinations were developed, but on the other hand, weapons whose practical value was secondary to a certain curiosity or technical gimmick.

A large proportion of the pistol-knife & stiletto combinations produced between 1840 and 1880 were developed and produced in Great-Britain and in their that time colony India.
However, in the 19th century Belgian gun manufacturers copied a lot of this British models and sold them worldwide.

Since combination weapons are sought-after collector's items, they are now occasionally manufactured as fakes. Some are completely new, others are made by assembling historical edged weapon and pistol parts.

Collecting weapons has always been widespread in society. This led to the creation of combination weapons for actual use, exclusive present and decorative combination weapons that worked but were not necessarily intended for use. There were even weapons where the combination proved impractical and the usefulness fell short of that of the basic weapon without a secondary function.
There were also combinations that had the potential to endanger its user, for example when a pistol and a knife were combined and there was a risk that the flintlock weapon would discharge when handling the knife because the powder would trickle out of the pan, or that a percussion weapon would fire a shot accidentally. On the other hand, a bladed weapon that was poorly attached to a pistol could cause cuts to the shooter if used carelessly.

The variety of combination weapons is therefore enormous. The most common were combinations of firearms with a cutting, stabbing or striking weapon. In a man-to-man fight, they offered a unique advantage against an opponent armed only with a close combat weapon. Before this one could use his own weapon, he first had to overcome the distance at which the shooting function of the combined weapon was effective.

But afterwards, in most cases, the shooter had a close combat weapon with the combination that was inferior to those of his opponents.

Finally, a note on the naming of combination weapons.

The name always comes from the type of weapon that dominates or is fully developed, or whose use or usability is paramount. This is the part of the weapon that is fully functional even without the other. For example, a large knife with an attached pistol system or a pocket knife with an integrated pistol is a pistol-knife. On the other hand, a pistol with an attached folding or rigid blade is a knife-pistol.

Contents   page

1. introduction 6
2. multi shot and multi barrel weapons 8
3. bayonets 14

4. single shot military pistols with bayonets
      a) flintlock pistols with knife bayonet 28
      b) flintlock pistols with stiletto above the barrel 28
      c) percussion pistols with integrated knife 30
      d) percussion pistols with socket bayonet 32
      e) percussion pistols with fold out knife-bayonet 33
      f) percussion pistols with stiletto above the barrel 34

5. double barrel military pistols with bayonets
      a) flintlock pistols with stiletto underneeth the barrel 36
      b) percussion pistols with socket bayonet 37
      c) percussion pistols with stiletto-bayonet above the barrel 39
      d) flintlock pistols with bayonets on the side of the barrel 40

6. traveling and pocket pistols
      a) single shot flintlock pistols with stiletto underneeth the barrel 41
      b) single shot flintlock pistols with stiletto on the side of the barrel 50
      c) double barrel flintlock pistols with socket bayonet 53
      d) double barrel flintlock pistol with bayonet underneeth the barrel 53
      e) three barrel flintlock pistol with bayonet underneeth the barrel 55
      f) single shot percussion pistols with fixed bayonet 56
      g) single shot percussion pistols with stiletto underneeth the barrel 57
      h) single shot percussion pistols with stiletto on the side of the barrel 72

i) double barrel Lechjaufeux pistols with stiletto above the barrel ………………………………………………	**74**
j) flintlock blunderuss rifles with stiletto ………………	**76**
k) percussion blunderbuss rifles with stiletto …………	**77**

7. revolvers with bayonets
a) revolvers with integrated knife-bayonet ……………	**79**
b) revolver with attachable bayonets ………………….	**79**
c) revolvers with stiletto under the barrel ……………..	**81**
d) revolvers with knife-bayonet on the side of the barrel	**82**

8. pocket knifes with shooting device
a) pocket knifes with shooting device …………………	**83**
b) pocket knife with pistol and knuckle duster ………..	**87**
c) pocket knifes with pinfire cartridge revolvers system Lechaufeux …………………………………………….	**88**
d) pocket knifes with center fire cartridge rvolvers ……	**90**

9. poacher weapons
pistols that can be converted to rifles ………………….	**91**

10. big knifes with double barrel pistol
Doumont combinations ………………………………..	**97**

11. combat knifes and daggers with knuckle dusters ……..	**98**
12. knuckle dusters with stilettos …………………………	**103**
13. shooting knuckle dusters ………………………………	**104**
14. revolvers with knuckle dusters ……………………….	**106**
15. multifunctionel pistols and revolvers ………………….	**108**
16. rifle with hidden combat knife ………………………..	**115**
17. hunting hangers with shooting device ………………...	**116**
18. pistol swords …………………………………………….	**119**
19. pikes with shooting device …………………………….	**120**
20. axes with shooting device …………………………….	**121**
21. various combination weapons ………………………..	**125**
22. modern variants ……………………………………….	**129**
23. combinations with none-weapons …………………….	**131**
24. prestigious weapon combinations …………………….	**138**

1. Combination weapons:

Combination weapons are the technical combination of at least two weapons into one unit. In the original sense and in most cases, at least one of the combined systems is a firearm. But that doesn't have to be the case. In this sense a halberd is a combination of a cutting and a thrusting weapon.

picture 1

The head of a halberd

It combines three different types of weapons.
The halberd was primarily an infantry weapon for defense against cavalry soldiers. It had a 2 to 4m long wooden shaft. The point is a lance for frontal attacks, the broad blade is a slashing weapon for side attacks, the hooked blade was used to pull the rider off the horse as he rode past or to grab the horse by a leg and bring it down.
Despite that versatility, a few halberds were manufactured in the 18th century with a flintlock pistol integrated into the left and right sides of the head.

The purpose of a weapon combination is to increase its effectiveness for an attack but also to avoid being left defenseless if the capacity of a firearm is used up.

The latter case may have been decisive for the development of such weapons. Loading a firearm in the 17th/18th century was time-consuming and often impossible in combat. The enemy certainly did not wait until his opponent had calmly reloaded, but on the contrary, took advantage of his defenseless situation.

The accuracy of historical firearms was comparatively low, but the technology of war was developed accordingly. The opponents lay, kneeled, or stood facing each other at a distance, and each of the three rows coordinated the use of weapons so that at least one of the formations was always ready to fire during the reloading time of the others. The effect of firearms lay in the concentrated firepower and resulting danger potential for the enemy, which deterred them from attacking with edged weapons. The latter, however, was the main weapon of the infantry. As soon as the "powder was used up", the actual attack with melee weapons took place. Even in the Franco-Prussian war of 1870/71, army commanders judged firearms, especially pistols, as "excellent for making noise".

In order not to be defenceless against the enemy after using the predominantly single-shot-muzzle-loading rifle, riflemen also carried a sabre, sword, or similar sidearm. However, reaching for the sidearm meant discarding the now obstructive firearm.

Since this was by far the most expensive portable piece of equipment for the soldiers, it was logical to look for a way to continue using the rifle as an effective weapon even when the possibility of shooting with it was not longer available.

The systematic development of weapon combinations dates back to the 17th century. At this time, lances and spears were standard alongside striking weapons, swords, rapiers, sabers and daggers. Lance fencing was part of military training. It was therefore logical to add a bladed weapon to the rifle, which practically made it a kind of lance.

According to historical reports, the first combinations of rifles with stabbing weapons took place before the middle of the 17th century in the French city of Bayonne. The city's name thus gave its name to edged weapons that could be attached to or folded out of firearms.

In the period of the late historicism, in addition to replicas of knightly weapons and amor, functional combination weapons were again produced purely for decorative purposes. They differ from their historical models in the way, that they are machine-assisted made, have partly cast elements, and are less time-consuming produced.

But the time of useful combination weapons was not over, when cartridge weapons appeared. Guns as dagger and stiletto combinations have been manufactured up until the 1920s. Probably more as a status symbol, because the utility value of these models was lower than that of normal weapons of the same class. Well-known examples include pocket revolvers with switchblades, brass knuckles with knifes and the French Apache revolver with knife and brass knuckles.

2. multi shot and multi barrel guns

Ultimately, the combination of two singleshot rifles or pistols to form a double-barreled rifle or pistol is also a combination of two similar or identical weapons.

Strictly in this sense, a revolver could be considered as a combination weapon. Revolvers have evolved from revolving multi-barrel combinations, that developed to single barrel pistols with rotating chamber-block.

picture 1

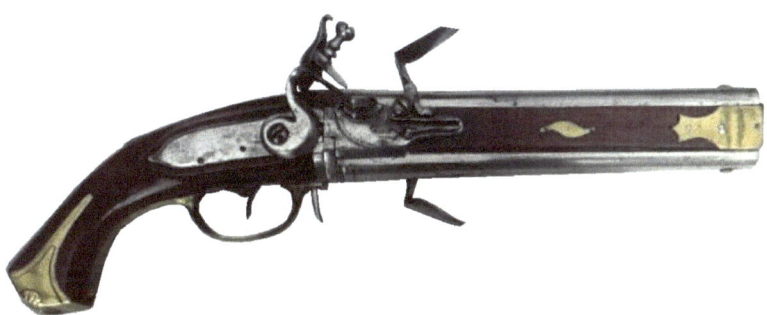

**Here a turn-barrel flintstone pistol from around 1800.
Collection Walter Gross**

picture 2

This is a hunting over-and-under combination from about 1860. It is a turn-barrel rifle with two barrels and one firing system, that serves both barrels, The barrel bundle can be turned, so that the upper (and possibly used) barrel comes down and the (loaded) barrel underneeth comes up. In some cases, both barrels have the identical caliber, in some, a different caliber or one barrel is using balls, the other shot.

picture 3

This oriental silver plated 3-barrel flintlock pistol is noteworthy. Usually flintlock turn-barrel weapons have a powder pan on each of their barrels. In this case, there is only one powder pan firmly attached to the grip above the rotating barrel bundles. It has a hole at its bottom through which when fired, the resulting jet flame penetrates downwards through the pan into the bunghole of the barrel below. After each shot, the barrels must be rotated by hand, so that the bunghole of the active barrel is located exactly under the hole in the powder pan.

picture 4

Side before: Four-Barrel flintlock pocket pistol, barrel length 6cm, total length 17,7cm, caliber 8,5mm, weight 382 grams

A lever on the left side of the weapon allows you to limit firing to the upper gallery or to all four barrels. This allows you to fire the two upper barrels simultaneously first and then switch to the two lower barrels. If you move the lever directly down, all four barrels will fire simultaneously.

picture 5

This is an extremely rare Japanese bundel revolver, a pistol with five hand rotated barrels, which dates back to the time when matchlocks, the earliest mechanized variant for firing a shot, were still used in Japan.
Collection Walter Gross.

However, the French Lemat revolver from 1856 is a combination weapon even without such a definition. In addition to the 9 cartridge chambers in caliber 10 mm or 8 mm that are located in the drum, the drum axis houses a shotgun barrel in caliber 16.8mm.

picture 6

https://commons.wikimedia.org/w/index.php?curid=583338

picture 7

Krieghoff over and under rifle shotgun ca. .16. rifle barrel 7x57, around 1980

Combination weapons have survived as hunting rifles in a modern form of the rifle above. Even the name combination rifle is still in use for those.

In the past, it was more common to combine two firearms into multi-barreled weapons, with each system firing the same type of ammunition or having the same calibre. Today, such weapons are more or less out of date, even among hunters. They have been replaced by repeating or self-loading weapons.

However, they have survived as a combination of a shotgun and a bullet system as a drilling, or even a quadruple or as over-and-under shotguns, or systems with barrels of different calibres, e.g. a small-calibre system and an medium-calibre system

picture 8

Kerner drilling in the shotgun cal. 16 and rifle barrel 5.6x50R, around 1930

picture 9

Krieghoff Plus drilling, shotgun caliber 12, bullet caliber 30-06 and in the right shotgun barrel a cal. .222 Remington insert system. Around 1980/90.

After cartridge weapons became popular, only the hunting weapons above remained as combination weapons.

3. bayonets

picture 10 and 11

two historical plug bayonets

picture 12

German military rifle with inserted plug bayonet. This gives the weapon a total length of 177 cm and makes it possible to keep an attacker with a saber at a distance where he cannot use his weapon.

picture 13

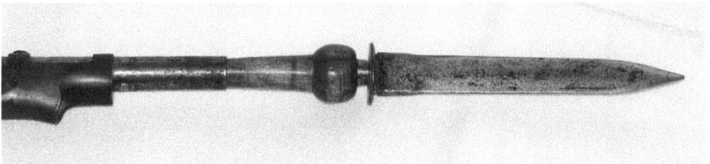

detail with muzzle and plug-bayonet

It is assumed that it were hunters who carved the wooden handles of daggers so that they could be inserted into the barrels of rifles, turning them into boar-lances with which they could fend off attacking animals after a missed shot.
Such daggers were then manufactured commercially as plug bayonets - mainly for hunting purposes. The idea was born.
The military use of plug bayonets is mentioned first in 1663 in Great Britain under the reign of King Charles II.

The disadvantage of plug bayonets was obvious. If the bayonet is fixed, it makes the weapon effective for thrusting, but on the other side it takes any chance for firing or trying to reload the muzzle-loader firearm.

It was therefore obvious that another method of attachment was needed. The simplest solution was to attach the bayonet with a sleeve around the barrel instead of inserting it into the barrel.
This could be done in such a way that no modification to the rifle itself was needed.
The idea was a short piece of tube that encloses the front barrel and to which a stabbing weapon was attached on the side. Usually, the front side of the end of the barrel was used to lock the socket behind it by turning it.

In 1689, the French army equipped its flintlock rifles with triangular bayonets.

In addition to the ability to reload the weapon with the bayonet attached, it was important that it could be mounted before the battle, meaning that the soldier was always equipped for both long range attacks and close combat.

But there were also bayonets with dagger or sword blades, such as the socket bayonet with a 51cm long blade, that was introduced in England around 1704 and could also be used as a sword. The socket served as a handle and was equipped with a guard like a sabre. A ball pommel could be inserted into the socket at the back and locked. Also interesting here is the sword developed as a model around 1790, in whose hilt rings were embedded that could be pushed around the barrel like a socket bayonet. This development continued and some countries introduced socket bayonets that could as well be used as swords.

In the 18th and early 19th century, almost all armies used such bayonets.

picture 14

This is a rare and possibly unique early socket bayonet.
It was pushed over the barrel muzzle and turned 90 degrees, so that it locks behind the front sight, but it blocks the barrel not others as a plug bayonet.
It combiners the advance of a socket bayonet with the disadvance of a plug bayonet. After mounted it, you can neather shoot nor load the weapon.
It is practically a transition model between plug bayonet and socket bayonet.
One could use the bayonet also as a dagger or mount it on a stick to create a spear or a lance. Length of the blade 15 cm.

early hunting bayonets

picture 15

Here a socket bayonet of the early 17th century. The socket has an expension joint that clamps it to the barrel. To prevent it, being pushed back further in combat, the blade swings slightly inward and is blocked by the muzzle.
Due to the close arrangement of the blade and the socket, the bayonet can also be used as a dagger, with the socket then serving as a handle.

picture 16

**flintlock hunting rifle with attached civilian socket bayonet.
2nd half of the 18th century.**
Collection Walter Gross

picture 17

Hunting socket bayonet with dagger blade, around 1800. The socket is made of brass.

picture 18

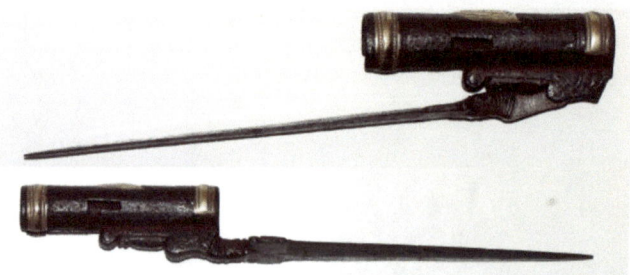

Here is a rare folding bayonet of the 17th/18th century. It bears the coat of arms of the French noble family Bonnin de la Bonniére de Beaumont.

Military bayonets of that time were longer, made of a single piece of steel and, for costs reason, usually did not have a knife blade but a triangular thrusting blade.

picture 19

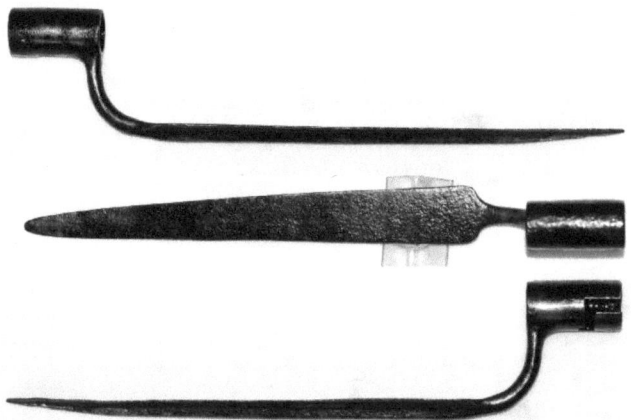

Here is a socket bayonet from around 1700 with it's usual fastening method. It is put over the muzzle so that the front sight slides into it's guide slot. At the stop, the bayonet is turned to the left (in this case, with others it may to be turned right) and than pushed back a little further in a subsequent slot guide.

picture 20

From around 1710, most military socket bayonets used a rotating ring as methode of securing the socket bayonet against unintentional sliding forward and twisting, that blocked the front rail behind the front rail.

The term bayonet lock, which is still commonly used in technology today, is based on this process.

picture 21

British perkussion rifle with attached socket bayonet. Total lenth is 1.92 meters

picture 22

This socket bayonet fits a muzzleloader pistol, length of the blade is 12.7cm, total length 19.3cm. It would fit a pistol with a barrel diameter of 22 mm at the muzzle.

picture 23

a socket bayonet for a double barrel rifle, triangle blade, around 185

picture 24

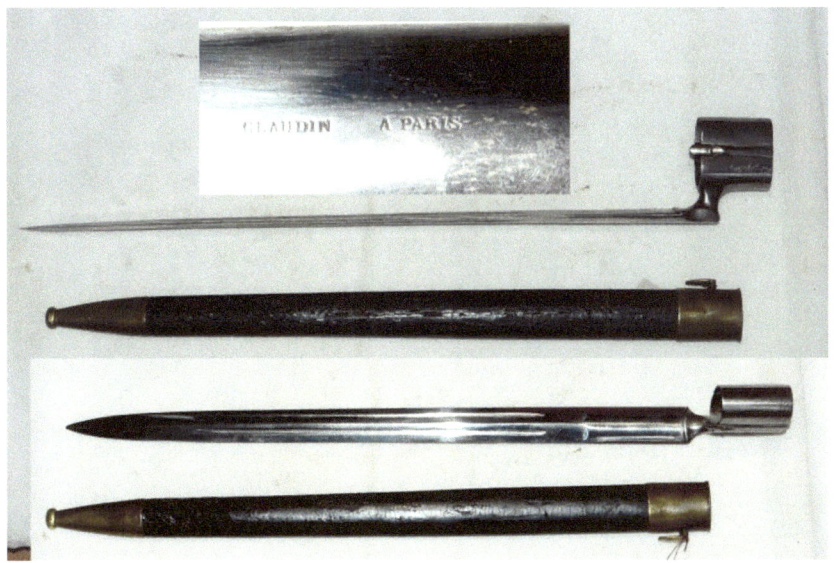

This is another socket bayonet for a double barrel rifle, the original sheet is dated 1854, the length of the blade is 46 cm

picture 25

Socket-bayonet for a double barrel muzzleloader pistol, length of blade 13.1 cm, total length 16.5cm. For barrels with 18mm outer diameter. The cal. of the belonging pistol around 14mm.

picture 26

French sabre bayonet model 1837, fitting an extra handle on it converts the socket bayonet into a slashing weapon.

picture 27

British East India Company bayonett, a sword socket bayonet, that can be used as a slashing weapon and bayonet as well, blade length 56 cm, 1840/1850

picture 28

Italian sword-bayonet for the carbine model 1868, blade length 44.9 cm, total length 57,9 cm

picture 29

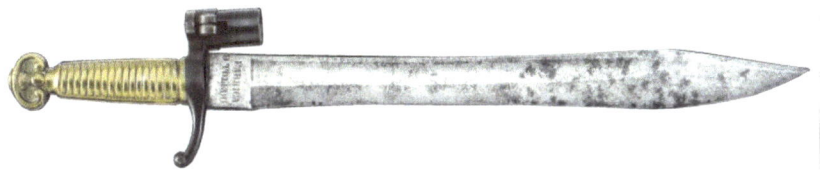

A Spanish bayonet model 1865, the blade length is 58cm.
It is an infantry sword with a socket for attaching it to the rifle barrel instead of the upper crossguard. Its use as a sidearm is therefore not restricted.

In Great Britain there was a transitional model, a socket bayonet with a sword blade, where a sabre-like handle could be inserted to the socket of the bayonet to use it as sword. However, all this socket bayonets had the disadvantage when used as edged weapon the handle was not at the same height as the blade, which proved to be a disadvantage in fencing.
Therefore, from 1800 onwards, a complete change of direction took place. The bayonet was no longer seen as a part of the rifle, but as an independent and fully fledged edged weapon, dagger, sword or saber.
It was given a proper handle, which included the option of attaching it to the muzzle of a rifle. However, this meant that the rifle had to be designed for this from the outset and that a suitable rail had to be present on the side or under its muzzle.
Now in combination with the rifle, the bayonet was a lance shaped stabbing weapon, but being used alone it was a full-fledged combat knife or a sabre or a sword, depending on shape and length of the blade.
In the First World War, the "mobile war" changed more and more into trench fighting. The long bayonets, swords and sabres proved to be a hindrance in the narrow trenches. Therefore, the armies involved in the war increasingly switched to short dagger bayonets.

picture 30

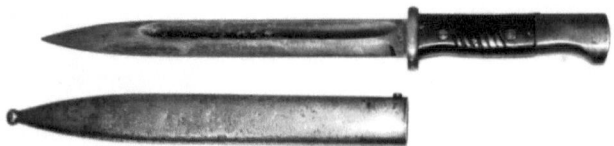

German knife bayonet 84/98 3rd type, WW2, a T-shaped slot at the top of the pommel, which was inserted into a rail under the muzzle of the rifle (here the German carbine 98k).
A locking button in the bayonet pommel ensures a secure hold.

As early as World war 1, when trench warfare with knives was common, German soldiers were able to acquire short trench daggers that could be attached to their rifles as bayonets. The most famous of these was the cranked Demag trench knife.

picture 31

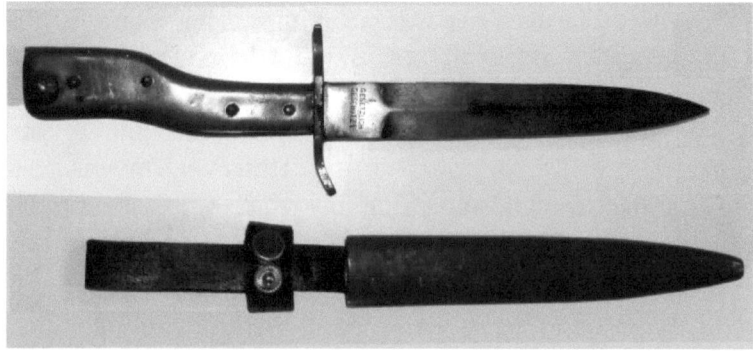

Military rifles of today still have the option of attaching bayonets, even though bayonet fights have certainly not taken place for hundred years.

The today's bayonets are combat knives and multifunctional tools as well, that also can be used as bayonets.

In the case of the Soviet Kalashnikov bayonet and some others can be used in connection with it`s sheet as a wire cutter.

picture 32

**Sovjet Combat knife and bayonet for the AK47 assult rifle.
In addition with it´s sheet, it works as a wire cutter.**

In Eastern Bloc countries, until the 1990s, and perhaps still today, there were wooden rifles in the shape ot the Kalashnikov AK47 with a spring loaded telescopic bayonet, which were used to train fencing with bayonet-equipped rifles.

picture 33

East-German (GDR) wooden AK47 trainings rifle with spring loaded telescopic trainig's bayonet

picture 34

The same training's rifle with bayonet attached

Rifle grenate launchers, which can be permanently attached to an assault rifle, are more commonly used in military applications today.

4. Single shot military pistols with bayonet

4. a) military flintlock pistols with knife bayonet

picture 35

This big British flintlock officer`s pistol from around 1800 is supported by a kind of plug bayonet. But in this case, the bayonet is not simply put into the muzzle, but in the muzzle is a winding, in which the bayonet is screwed in. The system has the disadvantage already described: with the bayonet mounted, the weapon can no longer be used as a firearm and cannot be loaded.

picture 36

belonging, 20,5 cm long bayonet

picture 37

The barrel length is 25,5cm, total length is 38,5 cm, total length with bayonet attached is 60 cm, it´s caliber is with 7.5 mm small for this period.
Weight without bayonet is 982 grams, with bayonet 1136 grams.

4. b) military flintlock pistols with stilettos above the barrel

picture 38

picture 39

The origin is unknown. The caliber is 17 mm, the barrel has plaster grooves. The barrel seams to be re-used several times, which can be seen from the fact that there are 4 different fastening notches for the bolt slot near its muzzle.
The barrel may have been once the front part of a rifle barrel. The stiletto is located above the barrel and can be released by a slide above the breech screw, where it automatically locks

when folded in direction of the grip. The flintstone weapon was made around 1800. Total length without the bayonet 28cm, the stiletto is 11.4 cm long, total weight is 793 grams.

picture 40

picture 41

Pictures 40 and 41 show the British tromblon flintlock pistol made arround 1820 by Goodwin & Son. Overall length without the bayonet is 31 cm. The calibre is 18mm.

4. c) military percussion pistols with integrated knife

picture 42

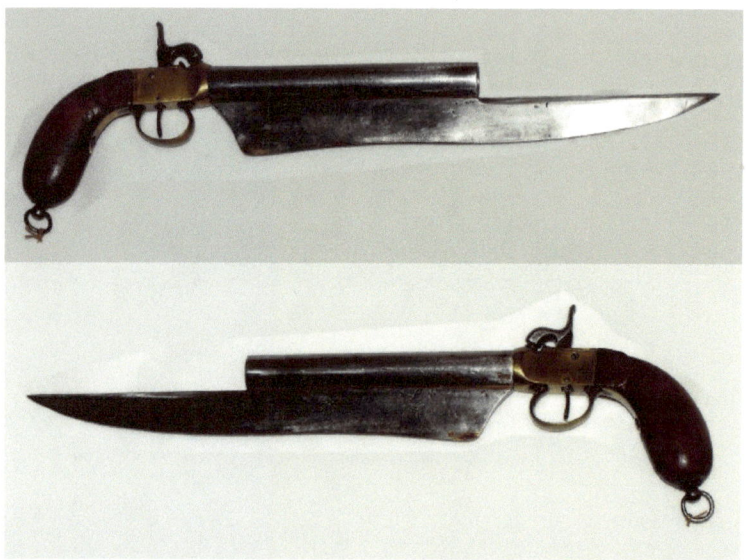

This weapon with a percussion box-lock has a 17.1 cm long barrel in 18mm caliber. The total length of the pistol without knife is 29.6cm. It's weight is 770 grams. The box is made of brass. A 32.2 cm long and heavy knife blade was attached to the unscrewable and unrivaled barrel with brass solder.

A rear sight is filed into the hammer. As a pistol, the weapon is a bit heavy and not good balanced. Due to the caliber and barrel length, it corresponds to a military pistol from around 1860 and probably dates from that time. The knife is also suitable for combat, but not to the same extent as a single combat knife, especially since it has no hand guard.

The lanyard hole indicates that it was also intended as a military weapon.

Because the knife blade protrudes far, loading the pistol is difficult and also carries a risk of injury. You need to be calm and able to concentrate. Reloading during combat is therefore slow wich proves to be a major disadvantage in a fight.

Overall, it is certainly an effective weapon, but it does not reach the effect as a single combat knife and a single percussion pistol.

The construction is very simple and there are probably many weapons of this type.

The brass box was cast, which suggests that it was manufactured in large quantities.

However, there are neither a manufacturing mark nor a proof mark on the weapon. But based on its look, I suspect that it was made in Belgium.

The pistol was probably also produced in a standard version without a knife blade.

picture 43

Abbildung wikipedia: Neochichiri11
https://commons.wikimedia.org/w/index.php?curid=23607595

US Navy knife 1838 - Elgins patent of 1837, screwed on 29.2 cm long blade, a combination of a percussion pistol, cleaver and knuckle duster, total length 42cm, caliber .54. Its barrel length is 12.8cm.

Only 150 of these pistols were delivered to the US Navy for their South Sea expedition. It was the first officially introduced military handgun of the US Army.

In the Navy, such weapon combination make sense when both enemy ships are close together during boarding. While in one place there is not yet enough proximity to use a saber and the opponents shoot at each other at a very short distance, in another place the ship can already collide and board opponents and attack the shooters from behind by surprise, so that when the weapon is fired there is no time and space to reach for the saber.

Original Elgin-Knife-Pistol combinations from 1838 are highly priced. But please mind, this weapon was later on produced for the civil market and those guns are much cheaper!

In addition, there are replicas of these!

4. d) military percussion pistols with attachable bayonets (socket bayonet)

picture 44

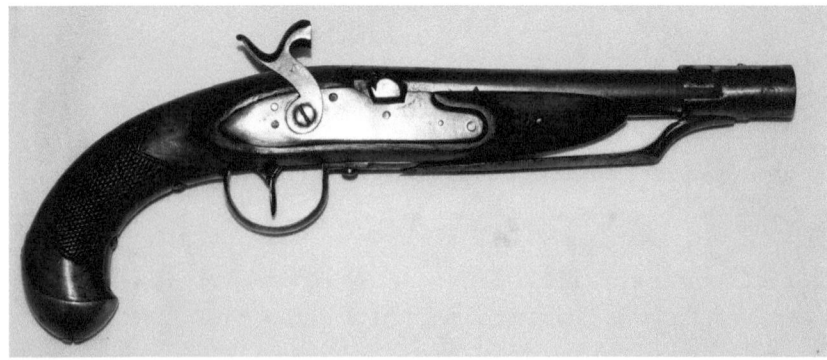

French military pistol around 1800, converted to percussion, with socket bayonet in non-effective position, barrel length 21,4cm, total length 38cm, the calibre is 17,5 mm, weight 1,14 kg

Picture 45

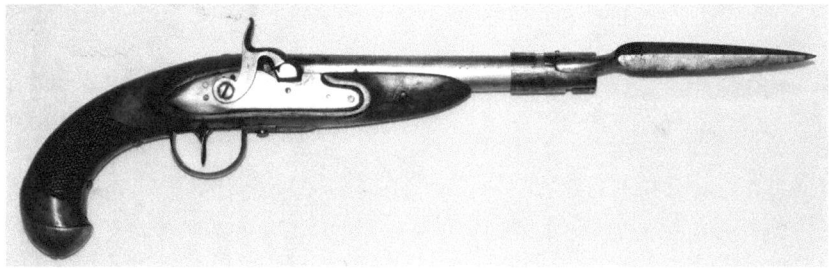

The same pistol with attached socket bayonet, making a total length of 52,7 cm. The bayonet of this pistol is the same as shown on picture 22.

4. e) military percussion pistols with fold out knife bayonet

picture 46

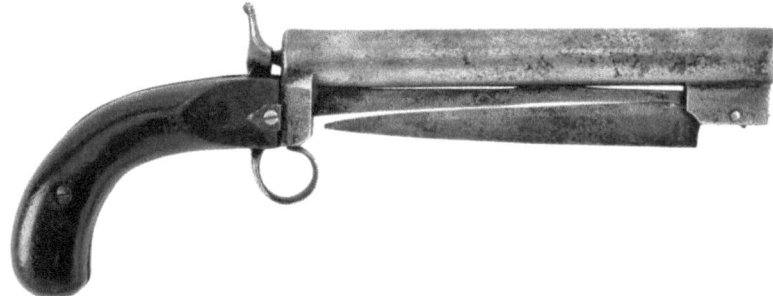

Percussion pistol with folding knife, cal. 17 mm, length unfolded 45 cm, folded 29cm.

picture 47

The same pistol with knife unfolded

4. f) military percussion pistols with stiletto bayonet above the barrel

picture 48

picture 49

Blair duck's beak pistol with stiletto, collection Walter Gross

This brass-barreled naval pistol from Blair in London was built around 1800 as a flintlock pistol, but was converted to percussion around 1840/50. It is a tromblone weapon in which the funnel-shaped muzzle is significantly wider than it is high, in order to achieve the greatest possible lateral dispersion of the shot used in close combat. This barrel shape is called a duck beak.

The barrel length of this weapon is 14.7cm, the total length 29.5cm. With the bayonet unfolded, the weapon is 38.5cm long.

The caliber of the rear cylindrical part is approx. 15mm, the muzzle 4.3x2.3cm.

5. a) military flintlock double barrel pistols with stiletto underneeth the barrel

picture 50

A miquelette lock pistol Spanish or North-Italian around 1820

picture 51

the same pistol with unfolded bayonet.

5. b) Military double barrel percussion pistols with socket bayonet

picture 52

double barrel percussion pistole with socket bayonet (turned backwards)
collection Walter Gross

picture 53

the same pistol with the socket bayonet turned forward

This double-barreled weapon was originally a flintlock pistol made around 1780. It was converted to percussion around 1840. A socket bayonet was probably added on the same occasion. For this purpose, the forearm, which originally reached to the muzzle, was shortened to make room for the double-barrel socket. The bayonet is normally attached with the blade to the handle and then rests above the barrel. To attach it, it is removed and turned 180 degrees and then protrudes forward under the barrel. The bayonet can be fixed in this position using the barrel latch that has been freed up by shortening the stock. The length of the barrel is 11.2cm, the total length is 24.5cm, the square bayonet is 12.6cm long, the caliber of the barrel is 15mm, the weight with bayonet is 893 grams.

5. c) military double barrel percussion pistols with stiletto above the barrels

picture 54

Double barrel percussion pistol with snap bayonet made by Jones Maker Birmingham, lenght without bayonet 21 cm, with unfoldet bayonet 33cm, calibre 12,5 mm.

picture 55

The same pistol with bayonet unfolded.

5. d) military double barrel percussion pistols with stilettos on the side of the barrel

picture 56

Italian officer's pistols made in Napoli, snap bayonet hinged on it's left side. Length 25,5 cm, calibre 16 mm.

picture 57

The same pistol with bayonet unfolded

6. traveling and pocket pistols

A large proportion of the pistol-knife & stiletto combinations produced between 1840 and 1880 were developed and produced in Great-Britain and in their that time colony India.
However, in the 19th century Belgian gun manufacturers copied a lot of this British models and sold them worldwide.
Since combination weapons are sought-after collector's items, they are now occasionally manufactured as fakes. Some are completely new, others are made by assembling historical edged weapon and pistol parts.

6. a) single shot flintlock tercerols with stilettos underneeth the barrels

It is assumed that this type of pistol was developed in England. This is indicated by the fact, that its first mention is a patent for pistols with spring bayonets, which the English gunsmith John Waters received in England in 1766.
These types of weapons got the most common combination weapons of the 18th and 19th century. Production took place mainly in Great Britain and as its copies in Belgium.
This type of pistol is also available as a double-barreled flintlock or percussion pistol. The multi-barreled percussion models are in some countries, as Germany, subject to the weapons-act and purchase permit and registration is required.

Flintlock boxlock tercerols with attached triangular stiletto; in this widely used design, the triangular stiletto has a spring that allows the blade to swing forward and lock unter the barrel after the trigger guard is pulled back.

picture 58

Flintlock pistol with spring bayonet by John Waters around 1800

This weapon is considered the prototype of the tercerol with spring bayonet

picture 59

a very similar pistol, made by the London gunmaker Wilson around 1800, length with bayonet folded 26,5 cm, with unfolded bayonet 39,8cm. The caliber is 13.8 mm

picture 60

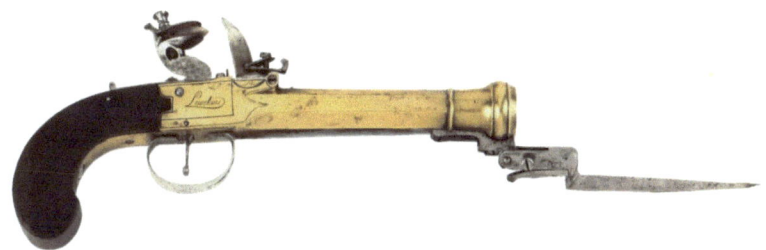

the same pistol with unfolded bayonet

picture 61

Pictures 60 and 61 show the British flintlock tercerol made by the gunsmith "Heeler" arround 1800/1820, it`s calibre is 14mm and the total length with unfolded bayonet is 44 cm.
The bayonet is not spring loaded and has to be folded out by hand.

picture 62

the same pistol as above with unfolded bayonet

picture 63

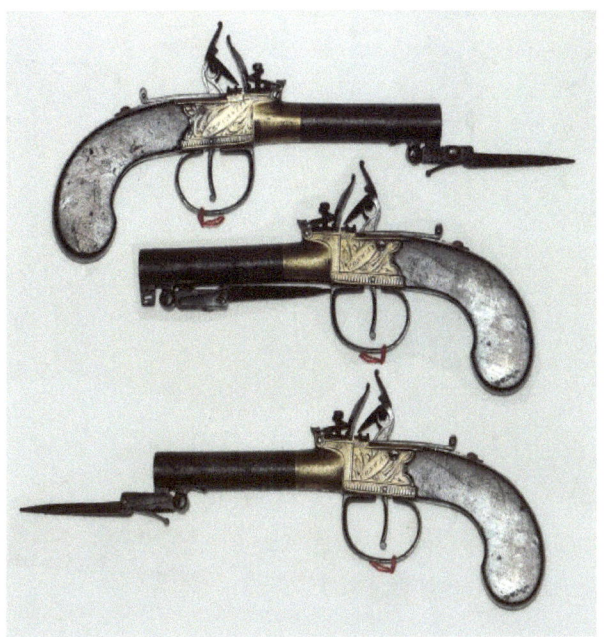

total length: 20cm, barrel 7,4cm, cal. 10.5 mm, weight 350 gr (+ missing upper lip of the cock)

The barrel length is 7,4 cm and the caliber is 10.5 mm. The barrel can be unscrewed. The weapon has a hammer safety. The weapon was made around 1840 in Great Britain by the Conway Newcastle company and bears British proof marks.
With the shown pistol, the upper hammer lip is missing. Without this, the weight is 350 grams. The pistol was made around 1840.

picture 64

British flintlock tercerol with bayonett, around 1820, total length 34cm, cal.. 12mm

picture 65

A small tromblon with snap bayonet, British made, total length with unfolded bayonet 23.5 cm, barrel length 14.5 cm, calibre 12 mm, weight 630 grams.

picture 66

The same pistol with unfolded bayonet.

picture 67

British flintlock pistol with snap bayonet around 1820, length with bayonet folded 24 cm, calibre 13mm

picture 68

The same pistol wit unfolded bayonet, length 31,5 cm

picture 69

A so called duckbeak pistol, others than a tromblon, the barrel wides at his front only to the sides, intended to by used as a riot gun. It was loaded with shot.
It is identical to the duckbeak pistol with flintlock, sprobly from the same maker and converted to percussion lock.
Total length 21 cm, basic calibre 15mm, weight unloded 430 grams.

Picture 70

the same pistol with unfolded bayonet

6. b) single shot flintlock tercerols with stilettos on one side of the barrels

picture 71

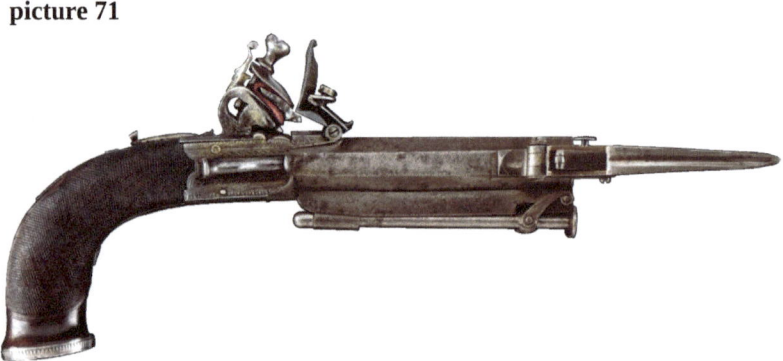

Pictur 70 and 71 show the flintlock pistol with the snap-bayonet on the right side of the barrel made 1910 by the British gunsmith James Haywood . It`s calibre is 11.5mm and the total length with unfolded bayonet is 28.5cm.

picture 72

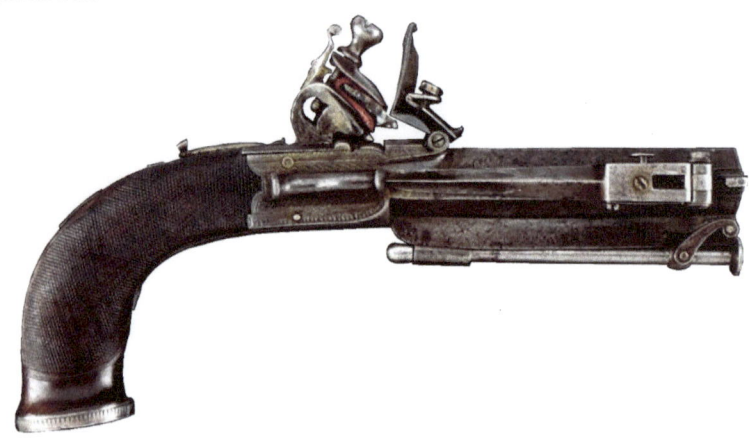

The same pistol with folded bajonet

picture 73

This flintlock pistol with a snap bayonet on it´s right side is made by the British gunmaker "Grice" of the city London. Box and barrel are of one pice of brass, barrel length is 13.5 cm, total length with folded bayonet is 25.4 cm, bayonet length is 9cm, the calibre is 15mm. The loading stick is of horn.

picture 74

here the same pistol with unfolded bayonet.

picture 75

A very similar pistol, made by the London gunmaker Bunney around 1820, length with folded bayonet 18 cm, calibre 12.8 mm.

picture 76

The same pistol with unfolded bayonet. Total length 29 cm.

6. c) double barrel flintlock tercerols with socket bayonet

Picture 77

Under-and-over pocket flintlockpistol with attached socket bayont, around 1820. Collection Harald Ehrenburg

6. d) double barrel flintlock tercerols with bayonet underneeth the barrel

picture 78

picture 79

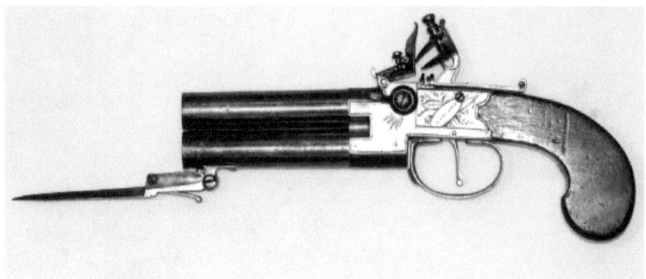

The weapon was made around 1830/40 in Great Britain by the gunsmith Hewson Piccadilly London. The lower or upper barrel can be activated using a switch lever on the right side of the weapon. The folding bayonet is activated by pulling back the trigger guard. The barrel length are 8.5 cm, the total length without bayonet is 22.8 cm, with bayonet 32 cm, the caliber is 12mm.

picture 80

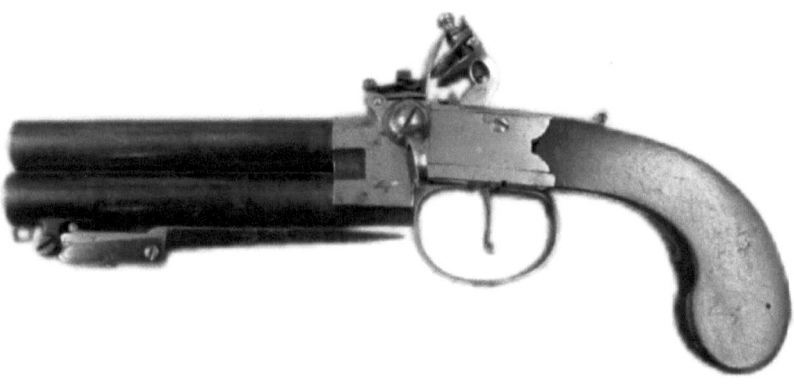

This is a very similar pistol as picture 122, but it does not show the maker and the fire selection switch is on the left side.

Comparable pistols are also available with a percussion lock.

6. e) three barrel flintlock tercerols with bayonet underneeth the barrel

picture 81

collection Albert Detmond

Three-barreled pistols are rare of themselves, with a stiletto bayonet even more rare.
The weapon was probably produced in Great Britain around 1830.

6. f) single shot percussion pistols with fixed knifes

picture 82

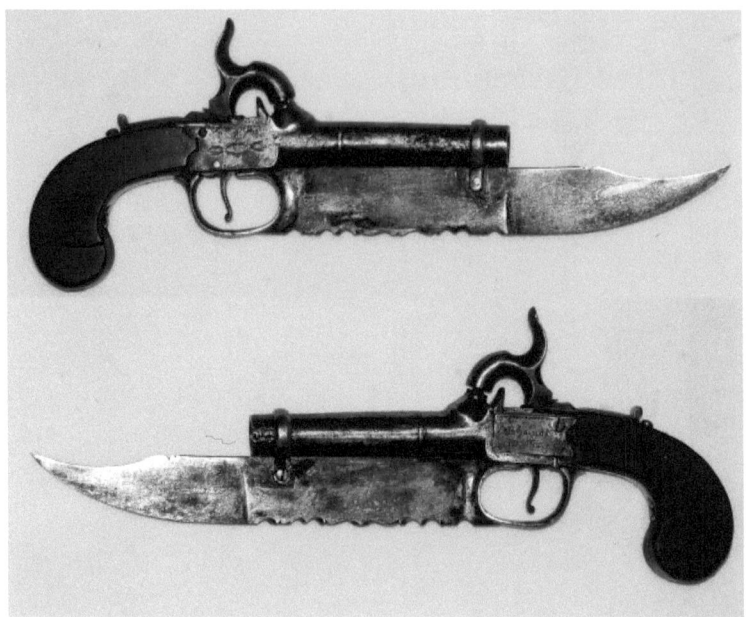

This weapon is a box-lock tercerol with percussion ignition. The 16.5 cm long bowie type blade ist welded to the trigger guard at the root. The blade is also stabilized at the muzzle with a barrel ring. It was manufactured - at least the basic weapon - by the British company 'Lorrot & Hughes' around 1860/70. The total length is 25.7 cm, the barrel length 8.5 cm, the weight 335 grams and the caliber 11mm.

The high-quality tercerol has a hammer saftey. The pistol was probably also offered without the blade.

6. g) single shot percussion tercerols with stilettos underneeth the barrels

picture 83

A flintlock tercerol by the British gunsmith Tacston with snap bayonet, converted to a percussion pistol.

picture 84

The same pistol with unfolded bayonet and cocked.

picture 85

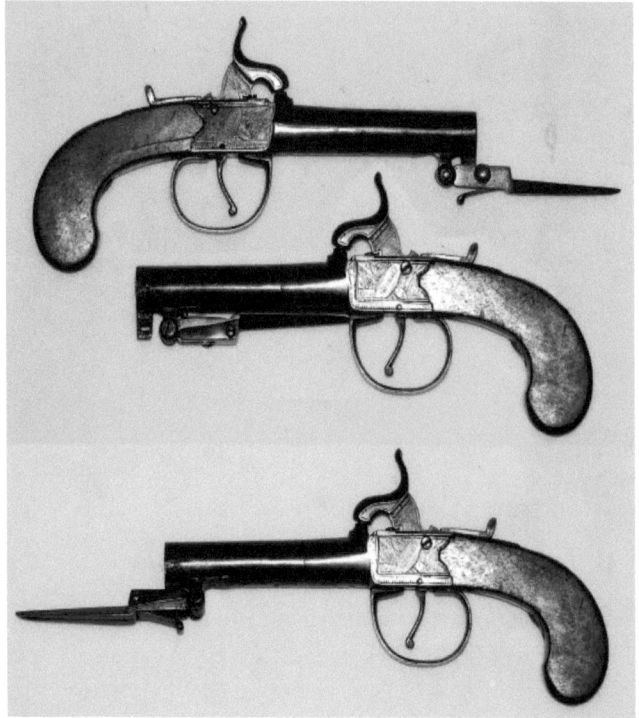

total length 19,1 cm, barrel 6 cm, cal 10.05 mm, weight 295 gr

Percussion box-lock tercerol with clipped-on triangular stiletto; in this widely used version, the triangular stiletto has a spring that allows the blade to swing forward and lock under the muzzle after the trigger is pulled back.
The total length with the stiletto folded in is 19.2 cm.
The barrel length is 8 cm and the caliber is 10.3 mm. The weight is 295 grams. The barrel can be unscrewed. The length of the stiletto is 6.5 cm. The weapon has a hammer safety.
The weapon was made around 1860 in Great Britain by the Southall London company and bears British proof marks.

picture 86

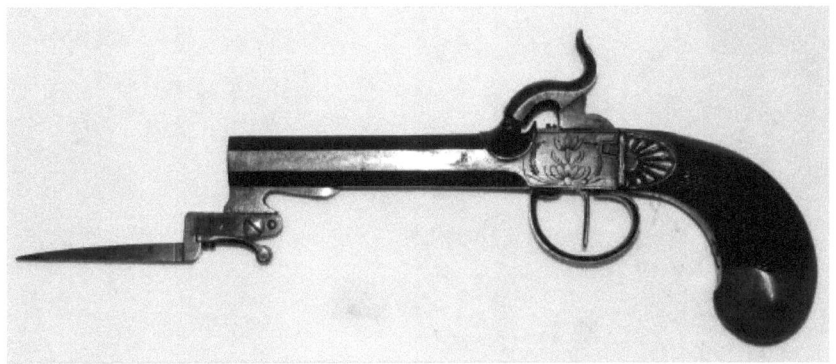

total length 21,2 cm, barrel 9cm, cal 12,4mm, weight 393 gr

Belgian tercerol with spring bayonet, box with floral engraving, triangular bayonet with 6.3cm long blade

picture 87

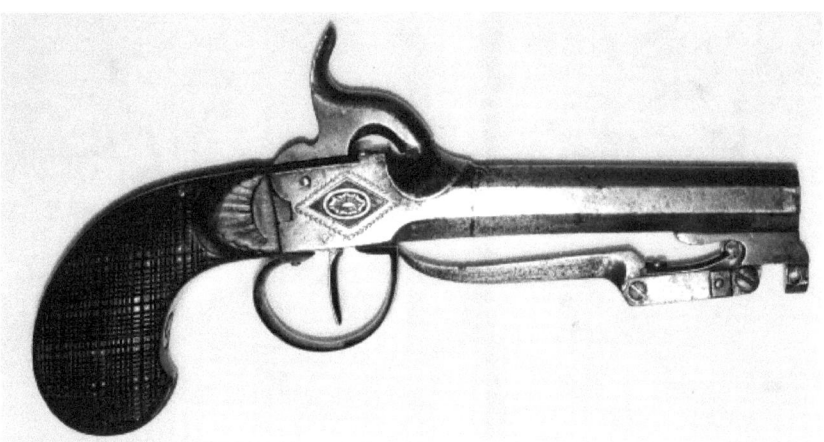

Belgian percussion tercerol with three edged spring bayonet

picture 88

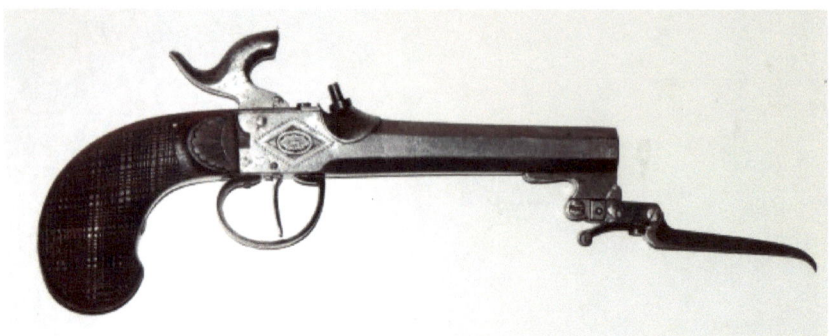

total length 21,7 cm, barrel 9 cm, cal. 12,5 mm, weight 445 gr

A fairly heavy tercerol with a spring bayonet. The downward curved blade is unusual, which is due to the relatively large distance from the barrel. The barrel length is 9.2cm, the total length 21.9cm, the caliber is 13mm, the weapon weighs 445 grams. The pistol has a Belgian black powder proof.

picture 89

total length 21,7cm, barrel 13cm, cal. 11mm, weight 330 gr

Here is another model of this type. You can see that the variety of this type of weapon is almost limitless.

Here the barrel is screwed to the box, its caliber is 10.8 mm, the barrel length is 10.3 cm, the total length with the bayonet folded in is 21.6 cm, with the bayonet unfolded it is 28.3 cm, the weapon does not have a proof mark, but probably comes from Belgium. The weight is 156 grams.

picture 90

total length 22,6 cm. barrel 11,1cm, cal. 12,5mm, weight 430 gr

Here one more Belgian tercerol with a three-edged spring bayonet.
The total length with the stiletto folded in is 22.5 cm.
The barrel length is 11.1 cm and the caliber is 12.7 mm. The weight is 243 grams. The barrel and the system box are made

from one piece. The total length with the stiletto folded out is 32.5 cm. The weapon has a Belgian black powder proof from around 1860.

picture 91

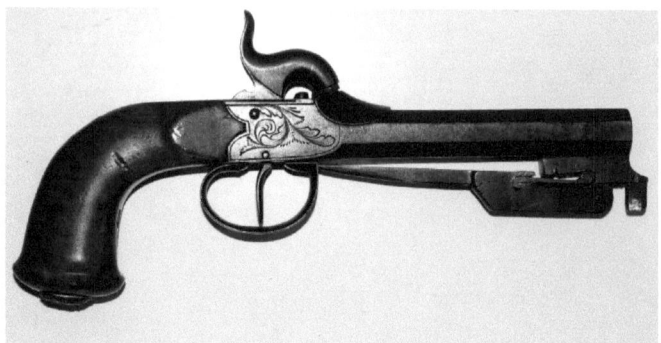

A Belgian percussion pistol with snap bayonet. There is no maker mentioned, but it is a good quality production and has a monogram shield on its but and a small box for primers. The bayonet does not really lock when cocked but is hold by a spring follower, so that it can be folded in without having to unlock it with a button as usual.

Picture 92

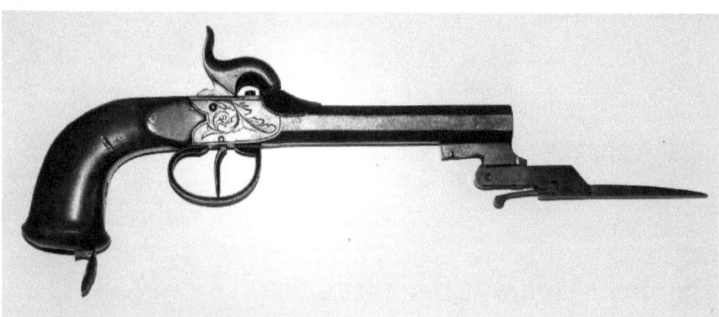

Here the same pistol with unfolded bayonet.

Barrel lenth is 11.3 cm, total length 23.8 cm, the length of the bayonet blade is 6.2cm, calibre is 11.5cm and weight 510 grams.
On the butt-end is a reservoir for caps.

Picture 93

total length 23,7cm, barrel 10,7, cal. 12mm, weight 364

Percussion box-lock tercerol with attached triangular stiletto. In this widely used version, the triangular stiletto has a spring that allows the blade to switch forward and lock unter the muzzle of the barrel after the trigger guard is pulled back.
The total length with unfolded stiletto is 23.5 cm.
The barrel length is 12.5 cm and the caliber is 12 mm. The length of the stiletto is 8.8 cm. The weight is 363 grams. The barrel cannot be unscrewed and has a Belgian proof mark on it. It was probably manufactured around 1860.

picture 94

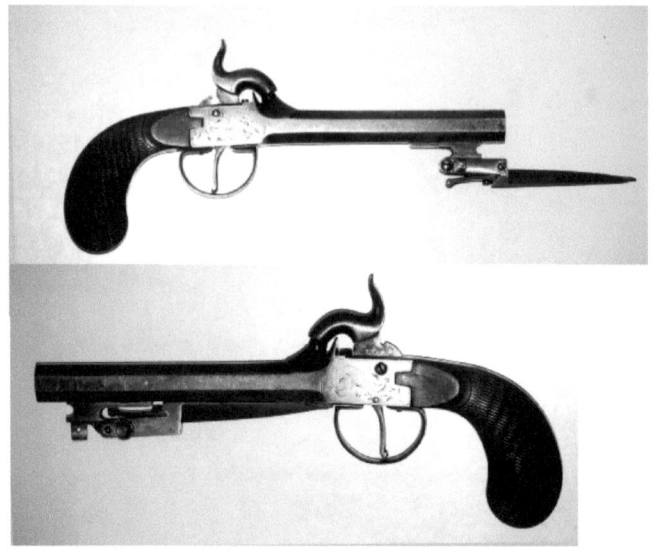

total length 24.1 cm, barrel 11.4, cal.11.7mm, weight 440 gr

The pistol has neither a manufacturer nor a proof mark.
The hammer is fish-shaped, the barrel and the lock box are forged from one piece, the lock box and the trigger guard are lightly engraved.
The caliber is 11.6 mm, Belgian proof. Barrel length without bayonet is 13.2 cm, total length 24 cm, with bayonet extended 31.2 cm. The weigth is 146 grams.

picture 95

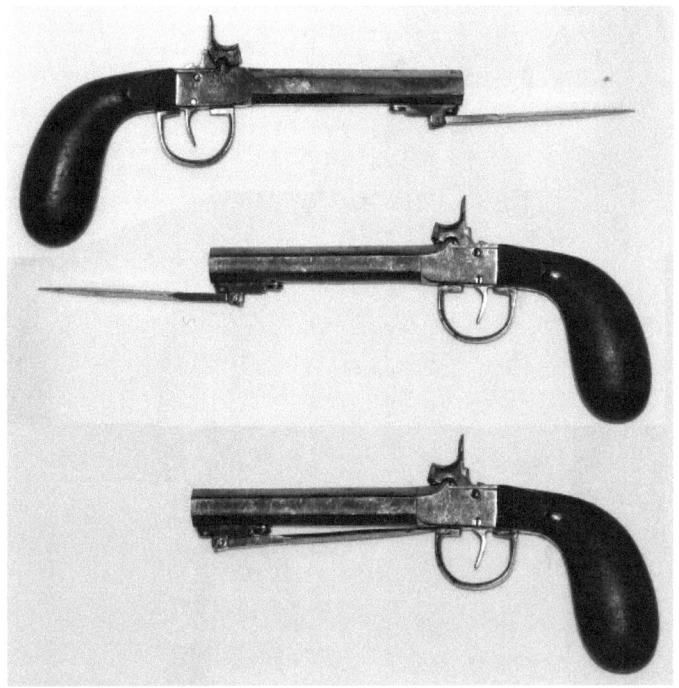

total length 24,2cm, barrel 11,9cm, cal. 14.5mm, weight 400gr

Chrom-plated percussion box-lock tercerol with clipped dagger stiletto, flat double-edged blade hinged to the front of the barrel, which locks under the trigger guard when folded.
The blade must be unfolded by hand with the trigger guard pulled back. There is no spring supporting this, but in unfolded position, there is a spring loaded catch to hold the blade in place.
There is a British proof mark on the weapon.
The non-screw-off barrel has a caliber of 14.5 mm and is 11.9 cm long. The total length of the weapon is 24.2 cm, plus the stiletto which is 10.1 cm long. The weight is 400 grams.
The hammer is offset slightly to the right so that there is space for a rear sight

picture 96

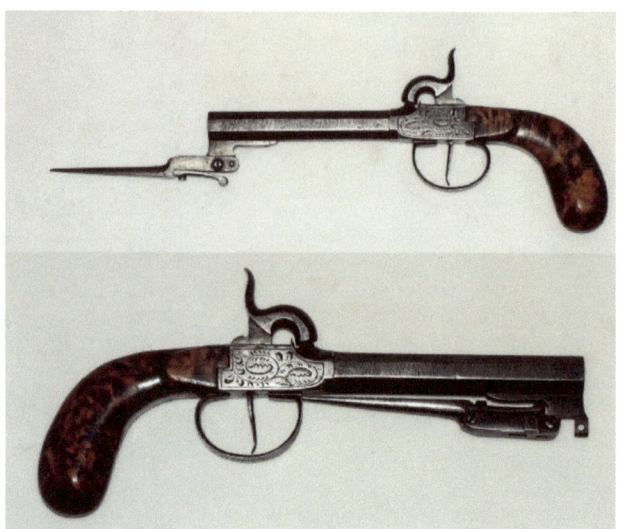

total length 32.6 cm, barrel 10.8, cal.11.4 mm, weight 385 gr

This pistol model corresponds to the pistol picture 69 and 70. The barrel has the manufacturer's stamp of the famous Belgium company Francotte. The barrel is screwed to the action box. The caliber is 11.4 mm.

The hammer is fish-shaped, the lock box and the trigger guard are engraved, the barrel has an embossed damask pattern. Length without bayonet 32,6 cm. The weapon weights 385 grams.

picture 97

Belgian percussion tercerol with snap bayonet, brass barrel and housing, barrel length 10,7 cm, Gesamtlänge , 22,4cm, caliber 11,7 mm, weight 442 grams

picture 98

The same pistol with unfolded bayonet.

picture 99

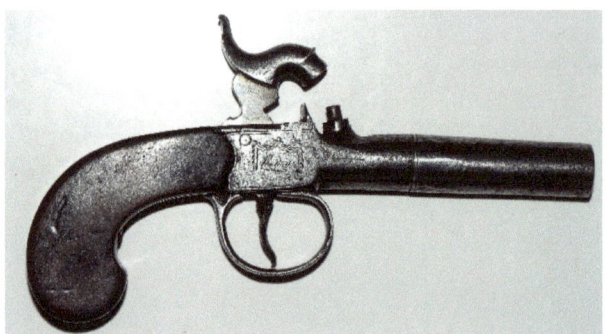

This little pistol made by the British gunmaker Blair is not a combination weapon, but it is a rare exemple of a tecerol, wich`s mechanic was prepared to be one. This 15.3 cm large pistol has a slidig trigger guard, that would impossible the gun, to lock and release a spring bayonet, being fixed under it´s barrel.

But actually there is no other indication, that there ever had been a bayonet on it, so the sliding quard never had any use.

picture 100

sliding mechanic

This picture show the sliding mechanic of the trigger guard. The spring has it's effect on the rear fixing of the trigger guard.

picture 101

This Belgian tercerol is very close to the shown weapon on picture 98.
It is the opposite story of the Blair tercerol above. A closer lock on it shows, that it once was equipped with a snap bayonet unterneeth the barrel, that was removed for any reason. The triggerguard can still be moved backwards, to release a snap bayonet, if fixed. Barrel length 9.3 cm, total length 22,6cm, calibre 13,7mm, weight 425 grams.

Picture 102

spring of sliding mechanic in front of it sliding triggerguard screw

triggerguard-slot for quiding an existing stiletto-bayonet

front end of the barrel, visible the remainings of the 2 screws, that had fixed the hinge of the bayonet

Opposit to the Blair pistol, the string for pushing the trigger guard in direction to the barrel end, has in this case it's effect on the front part of the guard.

Somewhat rarer than box-lock tercerols with central hammer and bayonet are those with side-hammer.
This one was made in Belgium.

picture 103

picture 104

Picture 104 shows a so called duckbeak pistol, others than a tromblon, the barrel wides at his front only to the sides, intended to by used as a riot gun. It was loaded with shot.
It is identical to the duckbeak pistol with flintlock shown at picture 68, so probly from the same maker and converted to percussion lock.
Total length 21 cm, basic calibre 15mm, weight unloded 430 grams.

6. h) single shot percussion tercerols with bayonet on the side of the barrel

picture 105

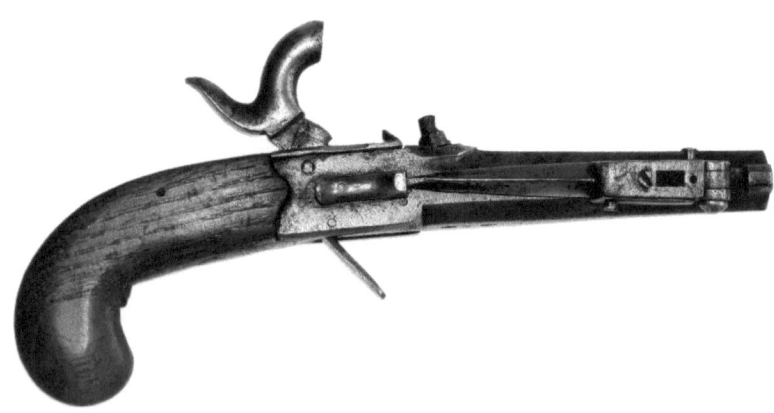

British tercerol with side bayonet folded, the barrel length is 8,9 cm, total length 18cm, caliber is 10,9mm, weight is 247 grams
The wooden grip is certainly new made.

picture 106

the same pistol with the bayonet unfolded

picture 107

this close up photo shows, that the pistol was build as a flintstone pistole around 1800 and converted to percussion around 1850.

6. i) Double barrel Lechaufeux-pistols with stiletto bayonet above the barrel

picture 108

Spanish Garrucha pistol, pin fire ignition, double-barreled break-action pistol in 11 mm Lechaufeux caliber, version with spring bayonet, manufactured end of the 19th century until World War 1. After World War 2 until at least the 1970s these weapons were produced in inferior quality for collectors.
Collection Walter Gross.

picture 109

picture 110

Picture 109 aand 110: French double barrel Lechaufeux pistol with flap bayonet. Pistol length 22 cm, with unfolded bayonet 30cm, cal.12 mm.

picture 111

the same pistol with fold out bayonet

6. j) flintlock blunderbuss rifles with stilettos

picture 112

Blunderbuss rifle (tromblon) with folding bayonet from around 1820.

Blunderbuss weapons are short firearms with a funnel-shaped widened muzzle that were loaded with shot, chopped lead or several round balls. They were used in close combat because their projectiles had a large dispersion and could thus take out several opponents with one shot.

In military terms, they were mainly used by ship crews, on a large scale in Great Britain and Denmark. In civil use, they were ideal weapons for carriage drivers,

As a single shot close combat weapon, it was obvious that they were also available with a folding bayonet.

6. k) percussion blunderbuss rifles with stilettos

picture 113

British Navy tromblon converted from flint lock to percussion. Hammer-spring in front of the hammer. Total length with folded bayonet 75.5 cm.
The basic calibre is 18.4 mm. Around 1800.

picture 114

The same tromblon with unfolded bayonet. Bayonet length 30,5 cm

picture 115

British Navy tromblon with percussion lock and hammer spring in the back of the hammer made by Blisset of Liverpool. Total length with folded bayonet 81 cm. The basic calibre is 16.8 mm.#Around 1850.

picture 116

The same Tromblon with unfolded bayonet. The bayonet length is 34,4cm.

7. revolver-bayonet combinations

7. a) revolver with integrated knife

picture 117

French knife-revolver from the Dumonthier company around 1860, cal. 9mm
collection Albert Detmond

b) revolver with attachable bayonet

Military Handguns with bayonets from the period after 1870 are rare. One of these exceptions is the Webley No1 Mk6 officer's revolver in calibre .455 for which the Prichard-Greener company developed a bayonet. This weapon was used by the British army between 1915 and 1947. During World War I , a bayonet and a shoulder stock could be purchased for this revolver, which turned the revolver to a short rifle.

picture 118

British Webley No1 Mk6 revolver with attached bayonet and shoulder stock

picture 119

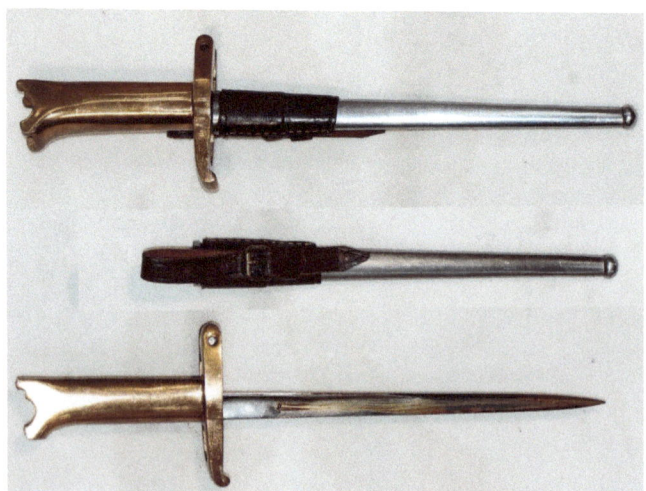

Prichard-Greener revolver bayonet for the Webley revolver No1MkVI

7. c) revolvers with stiletto under the barrel

picture 120

Lechaufeux revolver made by Miquel Basarte Eibar, Spain, around 1870, 6 shot calibre 12mm

picture 121

The same revolver with unfolded bayonet.

picture 122

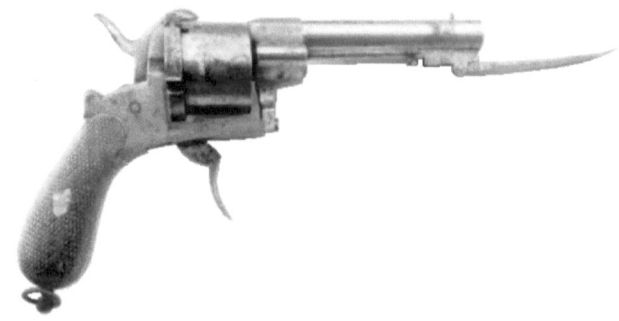

11 mm Lechaufeux pocket revolver, Belgium 1880 to 1910, under its muzzle is a fold out stiletto.
Collection Walter Gross.

7. d) revolvers with knife bayonet on the side of the barrel

picture 123

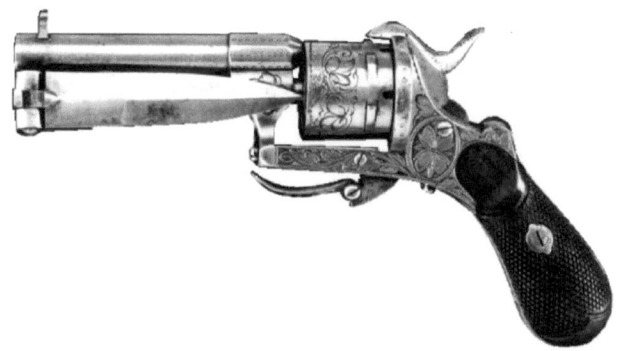

pinfire revolvers with folding knife

This Lechaufeux pocket revolver in caliber 7mm pinfire comes from Begium, where it was produced in the 1880-1900.

picture 124

The same revolver with unfolded bayonet

8. a) pocket knives with shooting device

picture 125

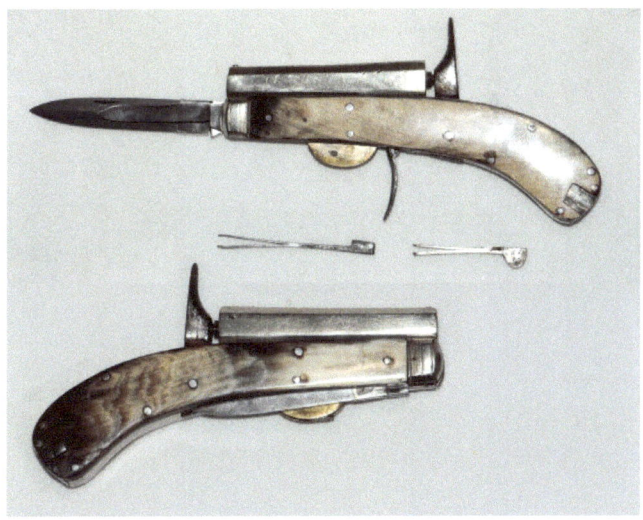

Unwin & Rodgers pistol pocket knife of 1861.
A pocket knife with large and small blade and a built in percussion pistol.
A compartment for bullets in the stock, a small bullet mould on the left and a pair of tweezers in the right.
The total length is 16.7 cm, the barrel length is 9.6 cm, the caliber 6,5 mm. Its weight is 457 grams.
The weapon has been copied identically or in a similar form by other companies.

picture 126

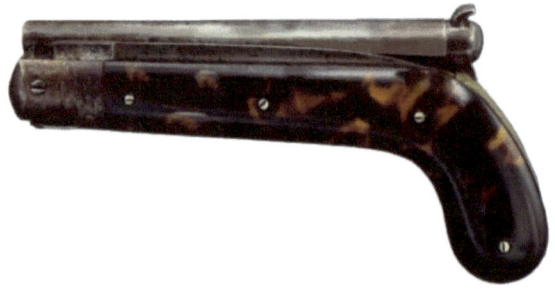

French pocket knife with pistol, marked Nogentaise of Langres, 8mm Perkussion, around 1850

picture 127

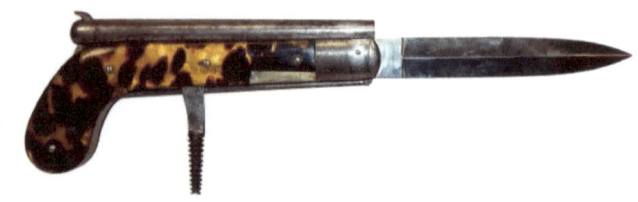

The same kind of pocket knife with perkussion pistol, but marked Lacouture, the cork screw serves as trigger.

Barrel length 12.5cm, total length unfolded 27cm.
collection Albert Detmond

picture 128

collection Albert Detmond

Here is another pistol pocket knife. The manufacturer is unknown. It probably comes from Italy or France.

picture 129

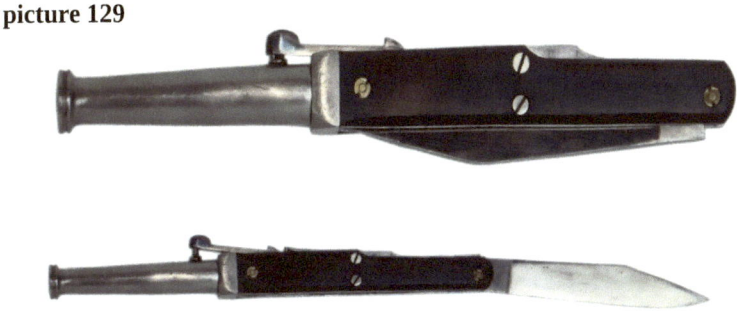

This pocket knife percussion pistol combination comes from France, probably from the knife-maker town Nogent. I estimate the period of manufacture to be the late nineteenth century

picture 130

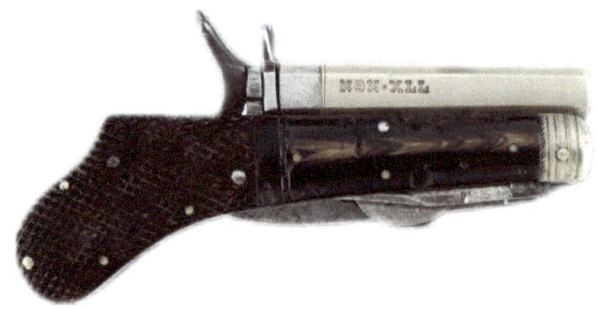

This British pocket knife from Unwin & Rodgers contains a pistol for rimfire cartridges in caliber 6mm and 9mm Flobert.

picture 132

the two bladed 9mm Flobert (rimfire) cartridge pocketknife pistole from Unwin & Rodgers.

The pocketknife pistols may also belong to the group of so-called cyclist's pistols, which were sold in large quantities a the end of the 19th century until World War 1 to fend off straying dogs.
collection Albert Detmond

8. b) pocket knives with pistol and knuckle duster

picture 133

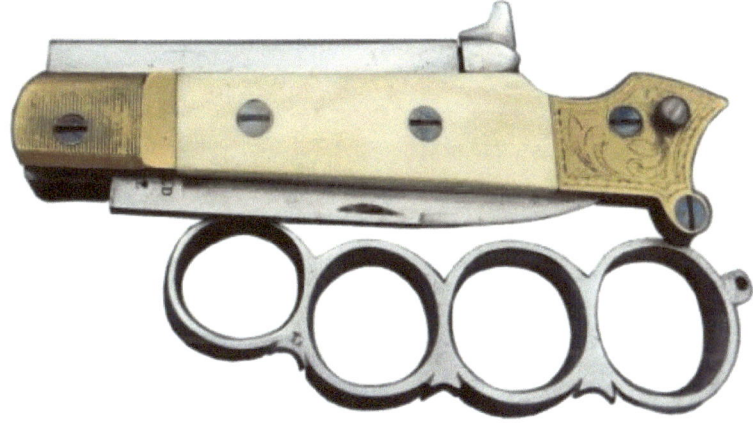

Belgian pocket knife pistol with knuckleduster, made by Arnould of Namour, percussion system with 8mm calibre

picture 134

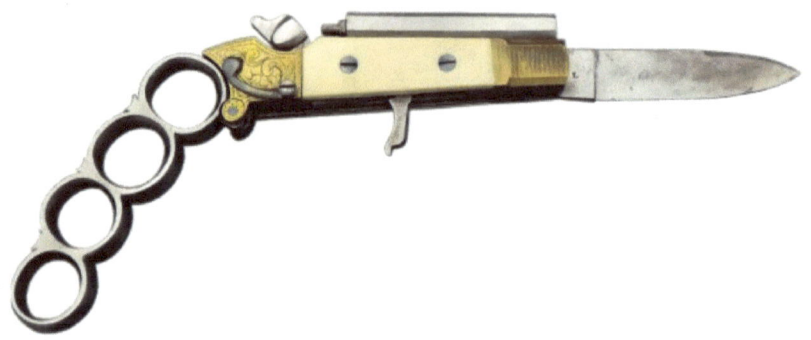

8. c) pocket knives with integrated Lechaufeux (pinfire) revolver

picture 135

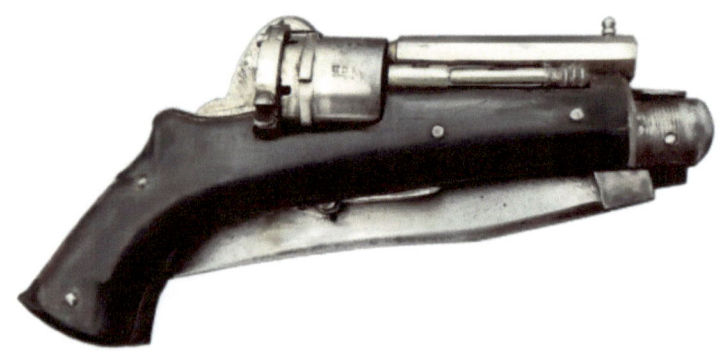

Pocketknife with 5mm Lechaufeux-(pin fire) revolver, French

picture 136

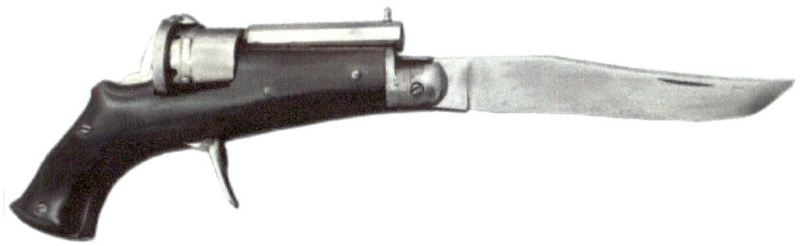

A similar French pocketknife revolver with unfolded blade. collection Albert Detmond

picture 137

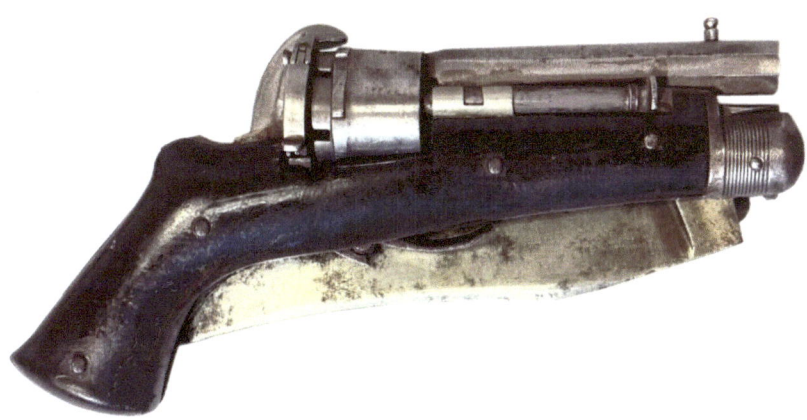

a Belgian Lechaufeux-revolver, cal. 7 mm, length with folded blade 23 cm, obviously a copy of the particular Fench model.

picture 138

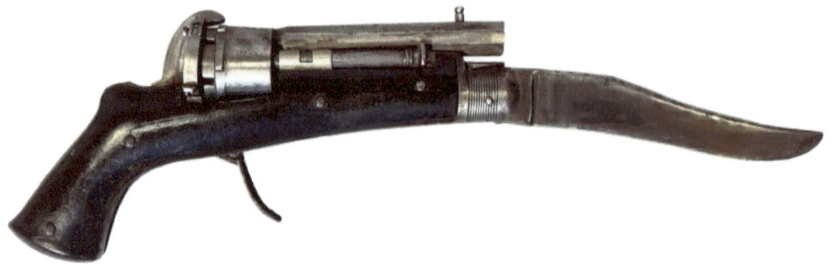

The same revolver as above with unfolded knife, effecting a total length of 37.5 cm

8. d) pocket knives with integrated revolver with center fire cartridges

picture 139

Belgian pocket-knife revolver cal. .320 centre fire, around 1920

picture 140

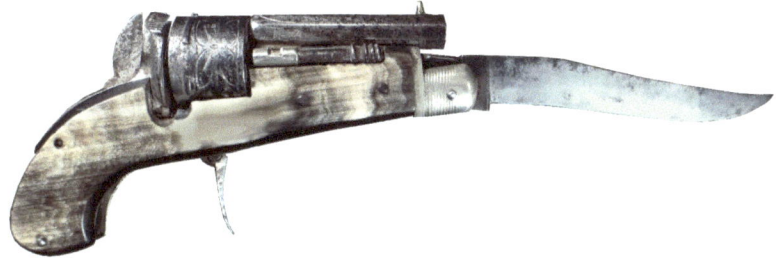

The same revolver with unfolded knife.

9. pistols, that can be converted to a rifle (poacher weapons)

picture 141

This flintlock poacher rifle was made by the gunsmith "Zeller", the middle part can be used as pistol.

collection Walter Gross

picture 142

picture 143

French percussion pistol around 1860. Manufactured in St. Etienne. So-called poacher's weapon. By screwing an extension barrel and attaching a stock, it can be turned into a rifle.
The caliber is 15.6 mm. The barrel length of the pistol is 12.4 cm. The total length of the pistol is 25.8 cm. As a rifle, the weapon is 126 cm long, of which the barrel makes up 89 cm.

picture 144

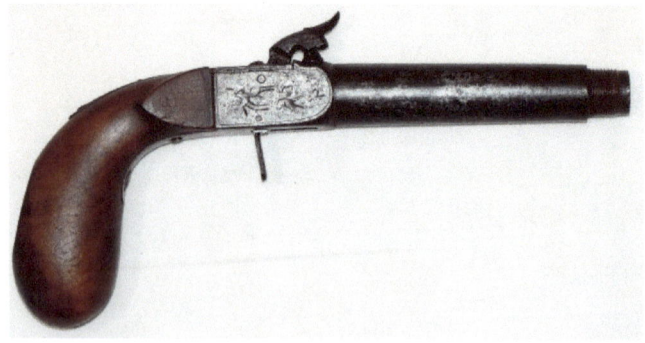

Here is an almost identical weapon, but there is neither a proof mark nor a manufacturer's mark on it. However, it certainly comes from France as well.

The barrel length is 15.8 cm, the total length is 27.8 cm and the caliber 15 mm.

picture 145

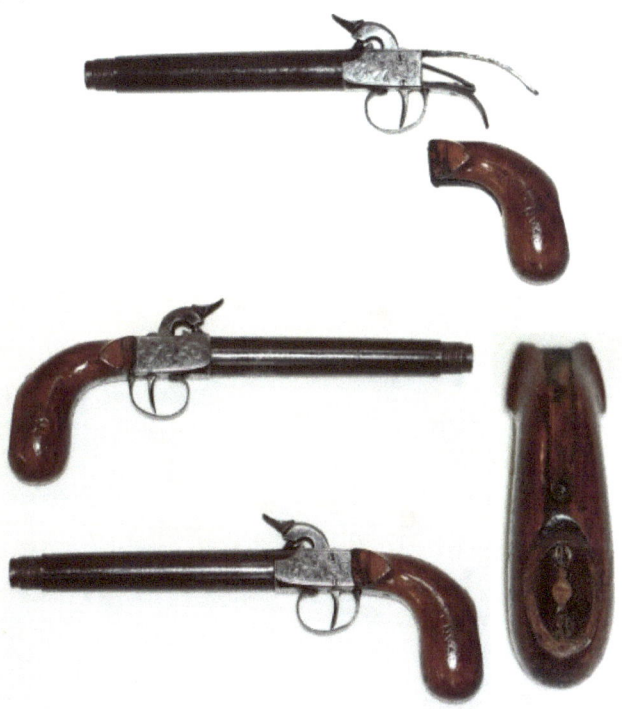

The belonging butt and the barrel extensions are missing here.

picture 146

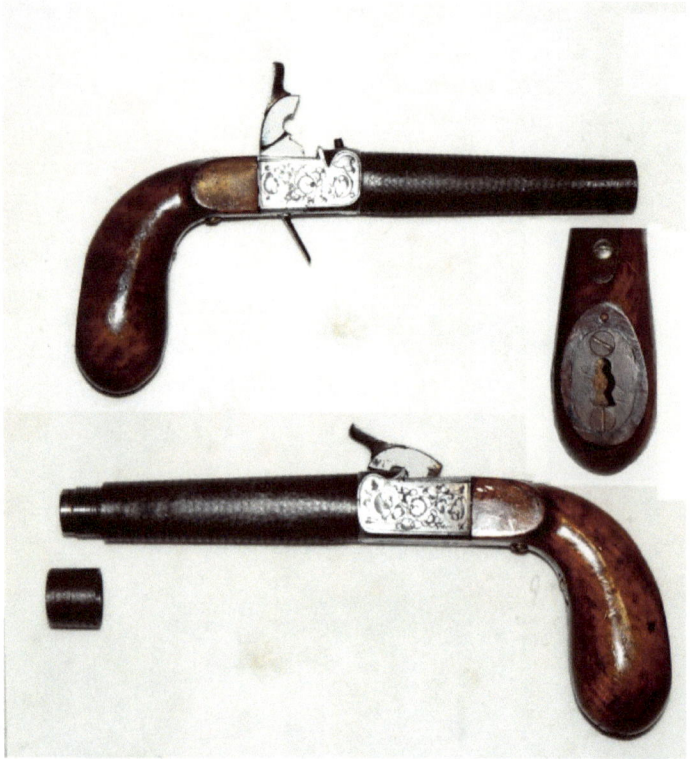

total length 24cm, barrel 12cm, cal. 14,6 mm, weight 575

Another weapon of this kind with unidentified proof mark, but certainly France. Fake Damascus type barrel of 12cm, caliber 15mm, total length 23,8 cm, with 572 grams it is pretty heavy, what comes from the massiv barrel.
Missing extention barrel and shoulder stock.

picture 147

percussion poacher rifle, maker unknown, collection Walter Gross

picture 148

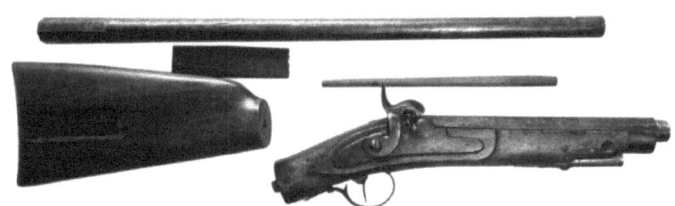

the same rifle as above disassembled, in this way, the middle part can be used as pistol

picture 149

This poacher rifle is missing it´s barrel extention. It was made around 1860/70 and there is no maker or proof mark on it. It was surely intended for poaching, what is shown by the fact, that the stock is painted black to avoid reflexes of light.
Total lenght as it is is 75cm, departed, the middle part is 49cm long and could be used as pistol if needed, The caliber is 15.5mm.

Picture 150

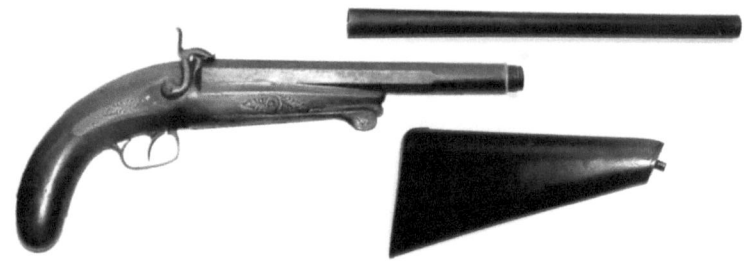

Lechaufeux pistol, cal. 12 mm, made in St. Etienne France

Picture 151

The same pistol assembled to a poacher rifle, total length 112 cm

10. big knifes with double barrel pistol

picture 152

collection Albert Detmond

The pistol knife from the French Doumonthier company is interesting. There is a barrel to the right and left blade. The two associated hammers form the crossguard when not cocked, the triggers fold out of the handle- caliber is 9mm, total length 32cm

picture 153

The Doumonthier-System, used by a different maker, blade 41cm, total length 55cm, cal.10,6 mm, not marked
collection Albert Detmond

11. Military combat knifes with knuckle dusters

In some countries, knuckledusters are prohibited weapons, what means, combinations with them are also not allowed in this countries.
Still 1917, the American Army introduced a combat knife with a saber-like handle, designed as a knuckleduster,

picture 154

American close combat dagger model 1917 with a 21cm long triangular blade, total length 33,5cm
Collection Piere Dubois

picture 155

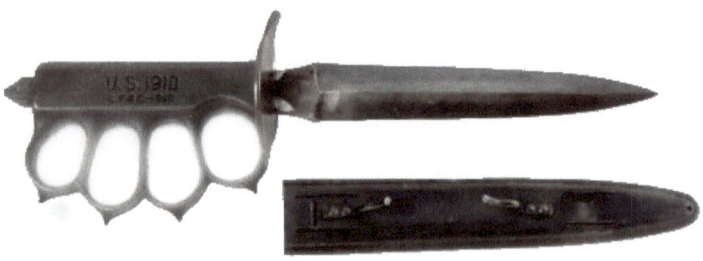

Foto cr: Carl Malamud - Flickr: M1918 Trench Knife, https://commons.wikimedia.org/w/index.php curid=16215828

Amerikan close combat dagger model 1918.

This model was a standard weapon of the US armed forces until 1943.

picture 156

This combat knife with knuckles was offered French soldiers to purchase for use at the front during World War I.
Collection Piere Dubois

picture 157

This also is a WW1 trench knife that French soldiers could purchase for warfare. It is more a knuckle duster with dagger.
Collection Piere Dubois

picture 158

French substitue dagger with knuggle bow model 1916
Collection Piere Dubois

picture 159

The British WW2 commando combat knife model BC41 with knuckles. It was issued before the Fairbairn-Sykes commando-dagger was introduced. The length of the blade is 13cm.
Collection Piere Dubois

picture 160

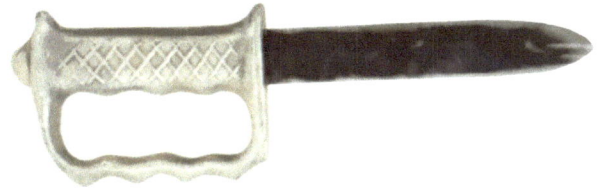

WW2 Combat knife of the 28th Maori Btl. New Zealand,
A combat knife with knuckleduster hilt.
Collection Piere Dubois

picture 161

British commando dagger for the Near-East and India Command, around 1950

picture 162

Australien Gregsteel knuckle knife model 1943, blade length 13.6 cm
Collection Piere Dubois

picture 163

Kukri fighting knife with knuckle duter, Nepal WW2

picture 164

A knuckle duster combined with 2 knifes, India late 19th to early 20th century, from point to point 29cm.

12. Knuckle dusters with stilettos

picture 165

British World war 1 Dudley push dagger with knuckleduster. Photo by user:geni, CC BY-SA 4.0, https://commons. wikimedia.org/w/ index.php? curid=52953327

picture 166

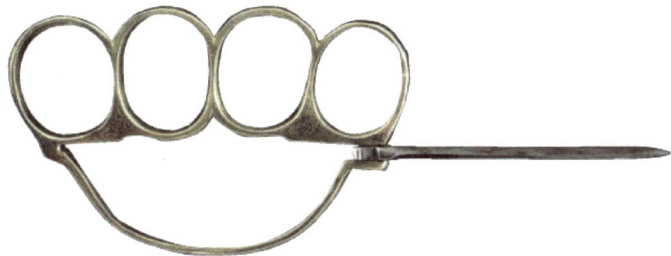

Here is a simple knuckle duster with a pivoting stiletto blade.
Collection Piere Dubois

13. Shooting knuckle dusters

picture 167

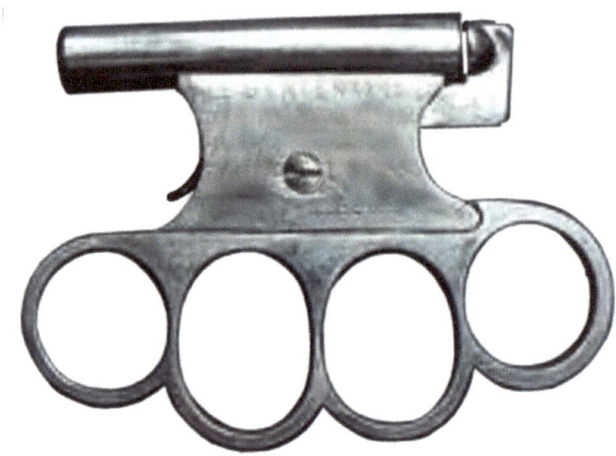

Knuckleduster with integrated single shot pistol cal. 5mm rimfire, probably French, manufacturer unknown.

picture 168

Another knuckleduster with 5mm single shot pistole 5mm cal. Probably French as well. This model exists in brass and steel.

Picture 169

French

kuckleduster with pistol, model „le Poilu", cal. 6 mm
rimfire, made after World War 1. Picture showing the face.

Picture 170

the back of the Poilu knuckle duster

Picture 171

This is another Poliu knuckleduster with the same face, but with the American crest on it's back. Probably made for export to the United States.

14. revolvers with knuckle dusters

picture 172

knuckleduster Reid's "My Friend", USA 1865

The "My Friend" revolver with knuckles function is, like the Apache revolver, one of the well-known combination weapons. However, the philosophy behind this weapon was different from the usual combinations of firearms with a stabbing or striking weapon, where the aim was not to be left defenseless after the last shot had been fired. Mr. Reid had the opposite idea, that one should not immediately defend oneself with a deadly weapon, but should first try to fend off the opponent with a less serious means, such as brass knuckles.

It's total length is 11cm. James Reid patented the weapon on December 26, 1865, but was only able to produce it after the patent for the "drilled through drum" expired. That was in 1869. Some more than 10,000 of the 7-shot revolver in .22 caliber were produced. There were also 5-shot versions in .32RF and .41RF caliber, although these are rarer. Later, even examples with a short barrel where made, but this was not a sales success, so they are extreme rare.

picture 173

Reid*s knuggle duster "New Model 32 - My Friend", the issue with cal. .32 rim fire. 3100 were produced between 1870 1nd 1882

15. multi functional pistols and revolvers

picture 174

picture 175

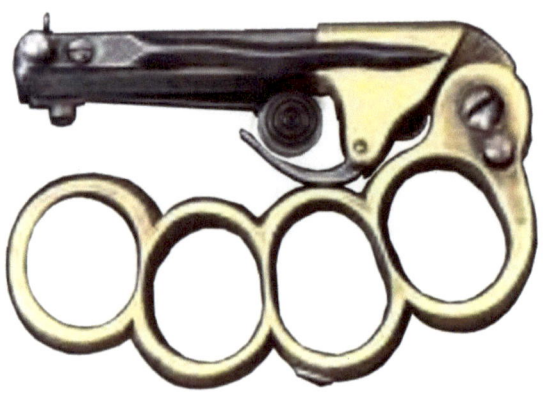

Picture 174 and 175 early Apache-type pistol with knuckleduster and stiletto, muzzle loader with under-hammer system, unknown French manufacture

The most known of this type of weapons ist the French Apache revolver and that caused many copies of it.

picture 176

picture: Apache revolver cr.: Latente Flickr 20
https://commons.wikimedia.org/w/index.php?curid=18750127

Apache revolver with unfolded knuckle duster and bayonet, brass frame and flamed blade. Unfolded 22 cm, folded 11 cm, weight 400 grams.
The weapon was introduced by Louis Dolne from Liege/Belgium at 1860 and produced until around 1870. I had 6 shot 7mm pin-fire ammunition.

Picture 177

Apache revolver folded and ready to use as knuckle duster

picture 178

Apache revolver, all steel, Belgian made, length unfolded 28cm, cal. 7mm pin firearm

picture 179

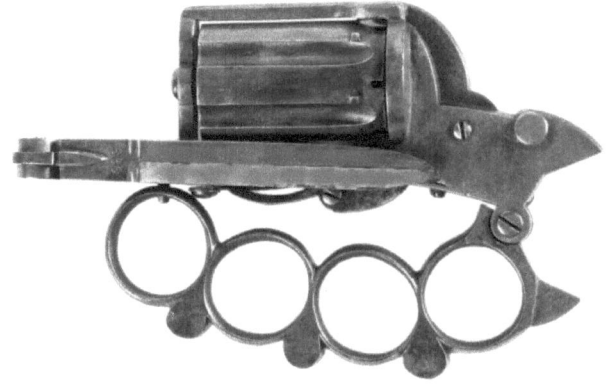

the same revolver folded to a length of 14cm.

picture 180

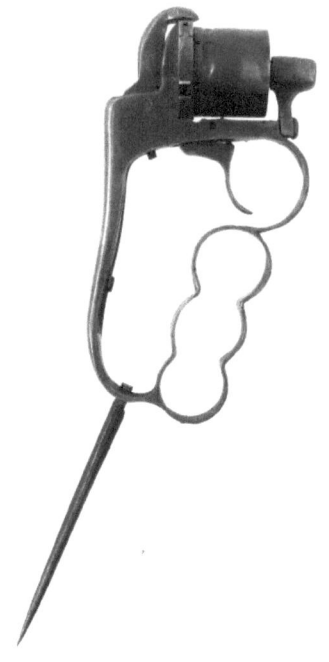

A variant of the Apache revolver of the French manufacturer Delhaxe, 6 shot cal. 7mm pin fire, around 1870, length 13cm, height without blade 12.5cm, with blade unfolded 19cm.
picture by the courtesy of IMA-USA.com, collectabe weapons

picture 181

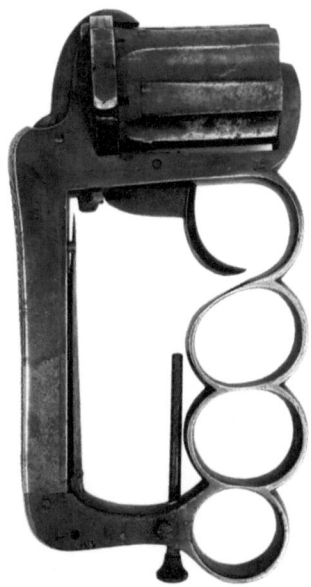

revolver with stiletto, cal. 7 mm Lechaufeux, made by the gun factory Delhaxe Liege Belgium

picture 182

the same Delhaxe revolver with folded out stiletto

picture 183

unidentified Lechaufeux revolver cal. 7 mm with knuckle duster

16. Rifle with dagger

picture 184

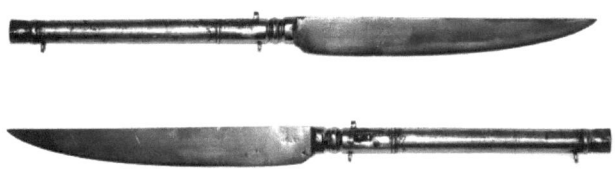

At first glance, it looks dangerous to shoot with this matchlock weapon. However, this weapon is missing an essential part. Id had a wooden buttstock in the shape of a rifle butt, the front of which served as the scabbard for the dagger. When the blade was in the stock, the weapon looked like a short rifle. Once the shooting options have been exhausted, the weapon can be pulled out of the stock and used as a cleaver.

The total length of the part shown here is 49cm, of which the barrel is approximately 22.5 cm and the blade 26.5cm. The caliber of the barrel is 11.3 mm. The weight of knife and barrel is 562 grams.

It was manufactured in India at the end of the 19th century.
Here the rifle-knife with a replica shaft and matchlock system.

picture 185

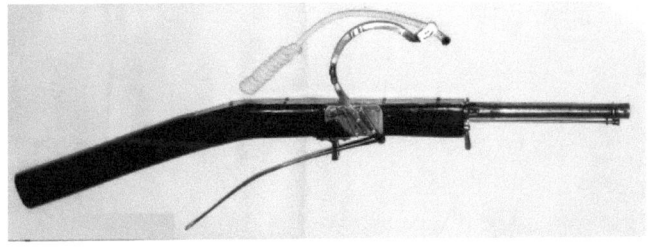

picture 186

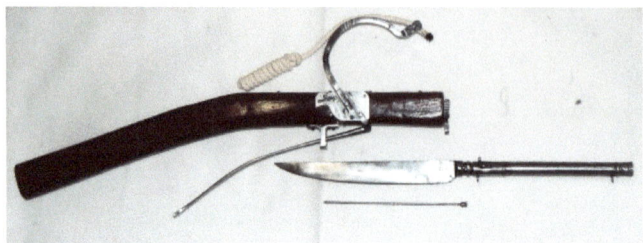

17. hunting hangers with shooting device

Hunting hangers with shooting devices are not uncommon. Their usefulness as a firearm was more theoretical. In addition to a certain status attribute, the idea was that a wounded large animal could suddenly attack if you tried to kill it with the blade. You could than immediately give it the finishing shot.
There are countless variants of shooting hunting hangers. However, due to their popularity, there are also many modern imitations.

picture 187

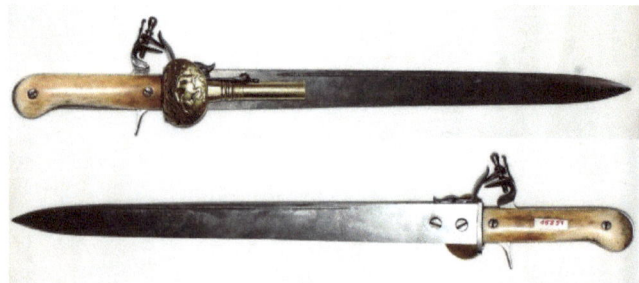

At first glance you would say that this is a hunting hanger with an integrated flintlock pistol. But it is the other way round, it is

a complete working flintlock pitol with a straight handle, with a hunting knife blade screwed onto the side of the barrel.
This weapon has a total length of 48.5cm. The blade length is 36.5 cm, the barrel of the flintlock weapon is 10.5 cm long and has a caliber of 10 mm. The weight is 541 grams. The barrel is made of brass, the handle shells are made of bone.

picture 188

There is neither a manufacturer's mark nor a proof mark on the pistol part.
It is probably a purely status weapon with little practical value. The screwed-on blade ist nowhere near resilient as that of a normal hunting hanger with a full-length tang, and at the same time the screw holes for the battery spring and blade do not only weaken the barrel, but the screws of the blade protrude slightly into the barrel, which does not allow a caliber-size bullet to be inserted deep enough to meet the powder chamber. So it may be, that this weapon was not intended for normal use, what would also explain the missing proof mark.

picture 189

Here is another hunting knife with a flintlock pistol. Its guard plate shows the French lily, France around 1800, total length 66,5 cm, barrel length 5,3 cm, blade 53 cm, calibre 9,5 mm, weigth 710 grams

Picture 190

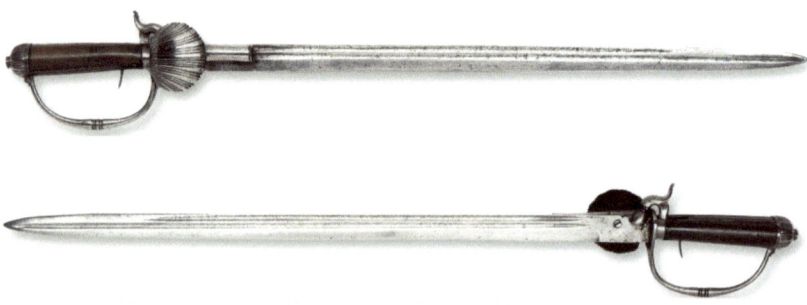

Hunting hanger with percussion lock, total length 76cm, blade length 61 cm, calibre 11mm, weigth 660 grams

18. Pistol swords

picture 191

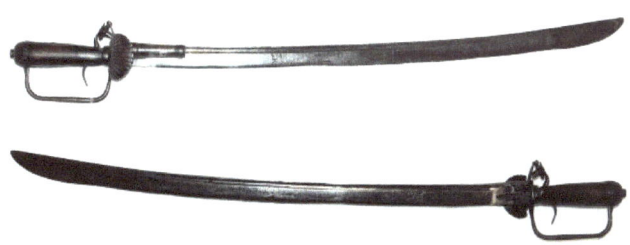

Saber with integrated flintlock pistol. Total length is 78 cm, blade length is 65.5 cm. The barrel is 9mm long and has the caliber 11mm. Possibly the weapon was made for hunters. It weights 714 grams. The blade is engraved on both sides with 'GR' and crown for Georg (3rd ?) of Great Britain.
Such a stamp is ofen found on weapons from India. Even though Georg the Third died in 1820 and Georg the Fourth in 1830, the sabre was probably manufactured at a later date.

picture 192

I beleave, this weapon was a fighting sword with percussion pistole before, that was shorted to a kind of hunting hanger. This is proved by the shape of the point and the fact, that it's fuller goes threw the point. Also the small calibre of 7.3mm of the pistol is is extreme unusual for hunting weapons as it has not the power to stop an attacking animal.

The grip is certainly not the one, with which the weapon was issued. The present blade length is 36,3 cm, the total length 50cm, the length of the barrel is 12,3 cm.

picture 193

Saber with integrated pinfire-revolver. Weapons like this were in use with officers in South-American States end of the 19th century.
Collection Albert Detmond

19. pikes with shooting device

picture 194

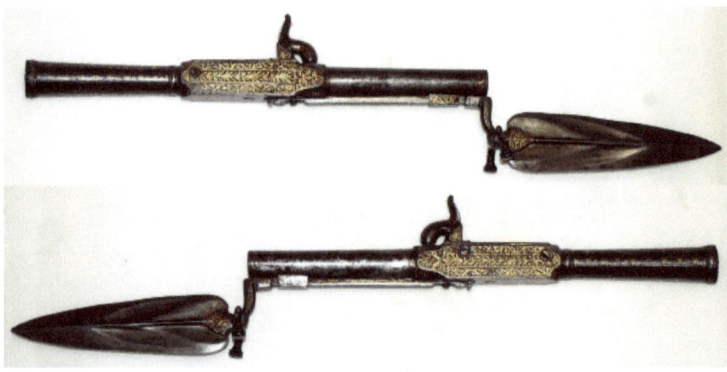

This is a hunting spear from India, according to the seller it was used for tiger hunting. The wooden shaft of the spear is inserted into the socket at the back. If the spear it pushed into a body, the shot of the pistol is fired automatically when the hammer was cocked. Around 1860/70.

System box of the percussion box lock, socket and tang of the spear blade with floral pattern as gold inlay. The total length is 45 cm, the barrel length is 10 cm, the caliber is 13.2 mm. The weight without the lance stick is 610 grams. there are no manufacturer's and proof marks on the weapon

20. axes with shooting device

picture 195

Small pistol axe with matchlock system (the hammer is missing), axe blade 8.5x13 cm, a barrel with match lock in the rear half of the iron handle. The axe head serves as a pistol grip, opposite the axe blade is a hammer head, the pan has a cover, total length 51.5 cm, barrel length 18cm, caliber 16.8 mm. Weight is 673 grams. India around 1850.

picture 196

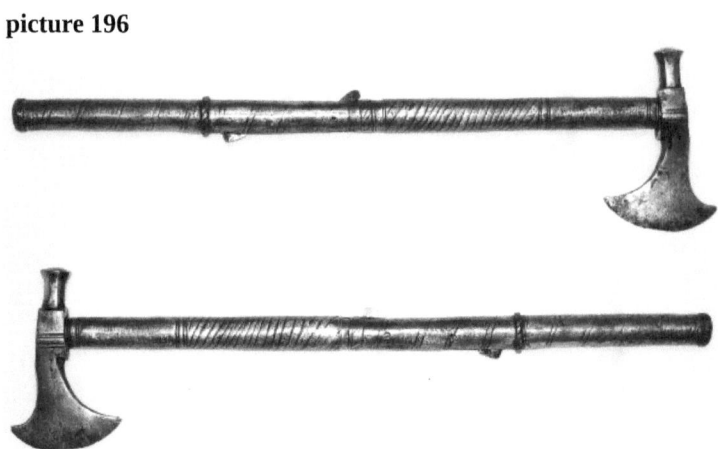

Pistol axe with matchlock, Idia around 1850, the barrel muzzle is on the side of the axe head, axe head 9x14 cm, opposite the cutting edge a war hammer head with a curved striking surface. The barrel caliber is 13.6 mm, the total length 52.3 cm, the barrel length approximately 23.5 cm. The weight is 1218 grams.

picture 197

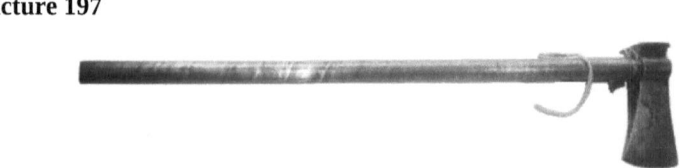

Axe pistol with matchlock, India 19th century, barrel lengt 63 cm, total length 67 cm. The caliber is 13 mm, the weight 2110 grams. The axe head is richly decorated with brass inlays.
Collection Walter Gross.

picture 198

axe with match lock, 16th century, Europe, calibre 12.8mm, length 99cm and weight 2.9 kg, shooing threw the head of the axe.

picture 199

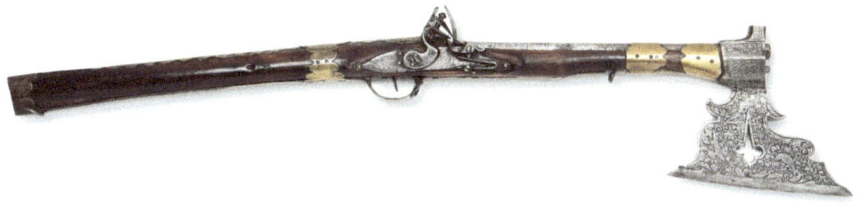

axe with flint lock, eastern Europe late 18th century, calibre 14,4 mm, length 94 cm, weigh 2.2 kg, shooting threw the head of the axe.

picture 200

This axe with flint lock has a total length of 85cm and a calibe of 11mm, it is shooting threw the head of the axe.

picture 201

European axe with flintlock, total length 85,5 cm, cal. 10.2 mm and weight 2.3 kg. It shoots threw the grip end

21. various combination weapons

picture 202

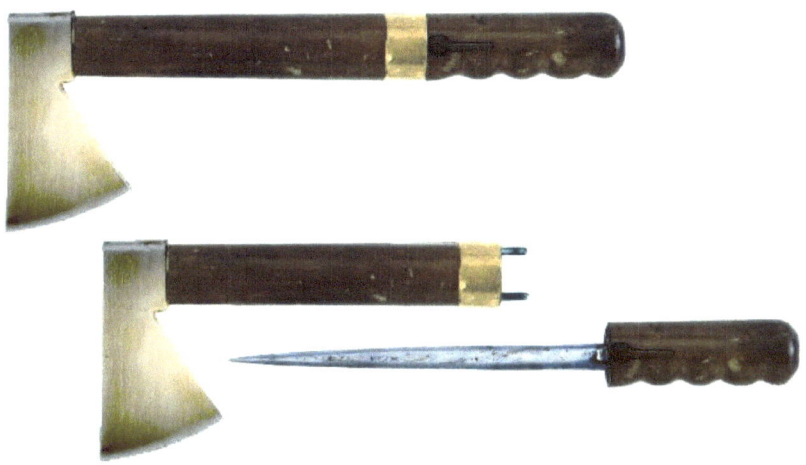

axe with hidden dagger, period and origin unknown

picture 203

Japanese Jitte, a striking weapon made of steel, with a hook of which the sword blade of an attacker can also be caught, Edo period. Here made as Teppo Jitte, which means that there is a barrel inside the stick, which in this case is ignited with a fuse (see bunghole under hook).
Collection Walter Gross

picture 204

Blowgun, probably Indonesia, 19th to early 20th century. A spear blade is attached to the mouth of the blowgun, which allows it to be used as a spear or lance.
Collection Walter Gross

picture 205

Extention sword

picture 206

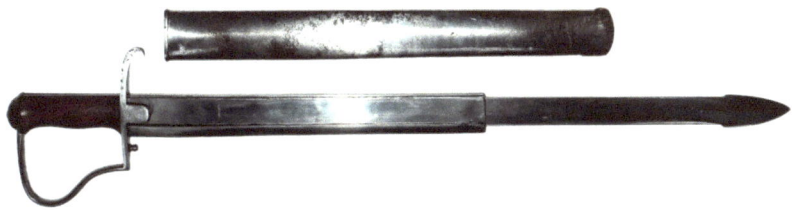

This sword is probably an Austrian forester's weapon. The hollow blade contains a second blade that springs forward when the button on the lower guard is pressed, extending the 47.5 cm blade to 68 cm. When pushed into the scabbard, the lock is automatically released, and the inserted blade locks again.

Chinese double swords

picture 207

To combination weapons, we also can count Chinese "sister-swords". They are a combination of two equal knifes or short swords, that are stored together in one sheet. For that reason, both identical edged weapons are totaly flat on one of their sides.

Places together, their contures appear as only one knife or short sword. That way, they are stored together in one sheet.

Being unsheeted an separated they can be unsed for two-handed fighting.

picture 208

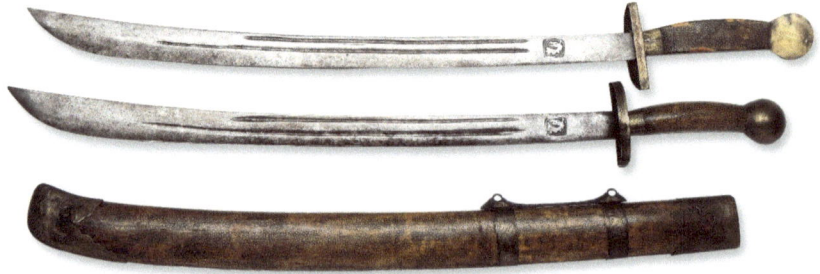

another pair of Chinese sister swords

picture 209

This madu from India is a combination of a defensive weapon and an offensive weapon. The shield is equipped with a stabbing weapon (animal horns) for close combat on the upper and lower edges. The weapon dates from the 19th century. The diameter of the shield is 17cm, the total width 51cm.
Collection Walter Gross

22. modern variants

The following exemples do not combine two different weapons, but are single weapons with dual use.

Hunting knife extents to the length of a hunting hanger.

picture 210

no maker given, but probably French

picture 211

French, maker Florinox Thiers

picture 212

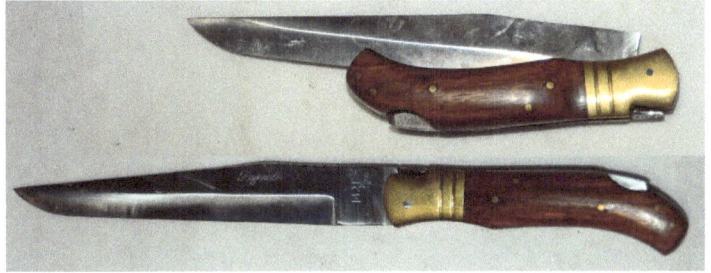

French, maker Laguiole

23. combinations with none-weapons

picture 213

Probably because of the limited availability of iron for ordinary citizens, scissors came into use in the Middle Ages, which were intende to be used also as a daggers. There were different models in use with different length, some even with crossguard.

Sissor closed, to be used as a dagger, blade length 8,8 cm, total length 18,2 cm

picture 214

the same sissor opend for regular use

It is not sure, if this is a European model from the middle ages. It ist possible, the the shown sissor was used around 1900 in India.

cane/walking stick combinations

picture 215

sword hidden in walking stick for self-defense collection Albert Detmond.

Picture 216

A pistol hidden in a walking stick for self-defense.

There are also versions with full-length barrel, but these are not really intended for quick defense against an attack. Because ot their length, they are more difficult and time-consuming to pull out of the stick but are more practical for targeted shot. They are therefore more likely to be used as poachers' weapons.
Collection Albert Detmond

picture 217

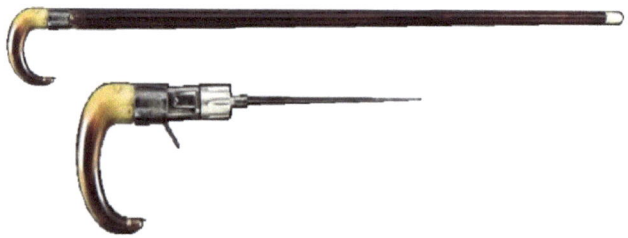

This is a bundel revolver with an additional stiletto blade, hidden in a walking stick.
Collection Albert Detmond

picture 218

Walking stick with a gun hidden inside, France. There is a very large variety of these weapons, as well as of cane swords.
Collection Albert Detmond

The pieces listed below are combinations of weapons with tools or common objects, whereby the weapons are not hidden but can be seen openly

picture 219

The weapon's characteristics are not hidden in this walking stick rifle. The barrel serves also as the stick and the firing system is a common and visible percussion boxlock, like it is used with tercerols. The model is probably French.
Collection Albert Detmon

picture 220

The Kalumet, a combination of a tomahawk and a pipe belongs to the combination objects. This has a total length of 50cm, the height of the blade including the pipe is 22.5cm. It origins from the Sioux Indians and was made in the 19th century.
Collection Walter Gross

picture 221

This is a combination of a war axe with an ancus (elephant goad), India from the 19th century. Length 57cm.

picture 222

collection Albert Detmond

It is conceivable that this book with its built-in percussion weapon was intended to provide a last, secret and surprising weapon for a robbery, even as the probability of ever having to use it was extremely low.

It was a kind of insurance against an unexpected attack in one's own home.

picture 223

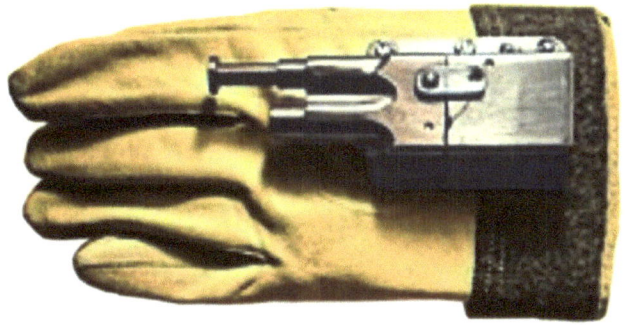

picture: Wikipedia

This glove pistol is a curiosity of the 20th century. **The weapon was concealed by a long-sleeved coat.** It was designed by the American company Sedgley from Philadelphia for the US Marine troops and Navy Intelligence for the Asian theater of war and was produced under the name 'hand firing mechanism Mk2' in a number of more than 50 and less than 200. It was a single-shot weapon and had a caliber of .38 Special. It was an assassination weapon used by the secret service to eliminate key opponents.

The protruding rod is the trigger. The shot was therefore fired by a clenched fist through direct contact with the victim's body.

24. prestigious weapons

picture 224

The following pistol case by the Aachen (German city) gunsmith 'J.G. Dachtine' from around 1780 contains a small all-metal flintlock pistol, as produced by Dactin in series and mostly in pairs. The grip contains a key the unscrew the barrel, so one of the pistols can serve the other.

In this case, instead of the second pistol, a 18.5 cm long gate key ist enclosed. With it a stiletto spike. When unscrewing the tip of the key, the spike can be fixed instead and turns the key into a stiletto.

The pistol length is only 16.5 cm.

The gate key with spike seams to be more a decorative combination, as there was certainly no fitting lock for it. The spike was not hidden or permanently mounted, but had to be laboriously screwed onto the key, when it's bearer was in a situation of danger.

Picture 225

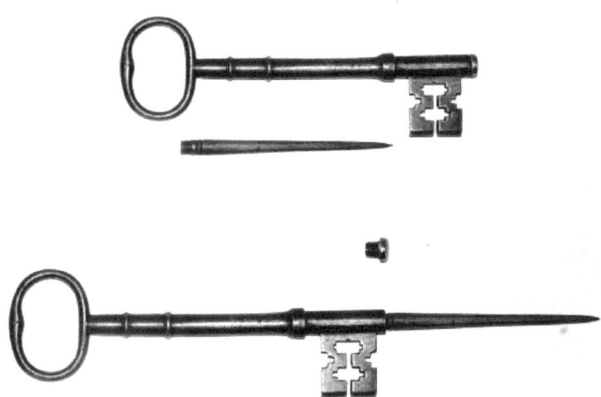

Carrying already mounted, needed the carry it in a scabbard, so there was no advantage over carrying a knife.
It therefore seems likely that ist was an honorary gift for a gatekeeper etc. .

picture 226

collection Albert Detmond

picture 227

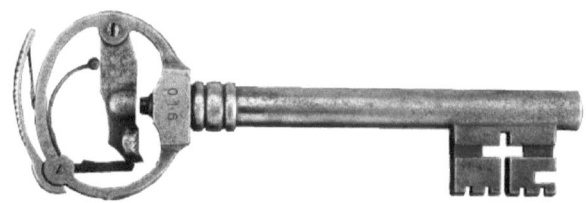

collection Albert Detmond
Gate keys as flintlock and percussion lock pistols.
The practical value of such weapons was low. Logically, a gate guard was armed in the traditional way. A key as a weapon was unwieldy, cumbersome and of little effect.
With the key with the percussion lock, there was even a risk of burns or splinter injuries from the exploding percussion cap when firing, which made it doubtful that its use was even intended,
I estimate that almost 100% of all these weapons date from the historicist period to the present day and were made exclusively for enthusiasts and collectors.

picture 228

This finger ring revolver was produced in a small series by the British jeweler John Smith of Lincoln in the 1920s. Its impact was probably quite limited due to the small powder charge and the small caliber of around 2mm. It loaded five pin-fire cartridges. It was therefore little more than a fashion accessory.

collection Albert Detmon

Combination weapons and tools reflect a time when versatility and resourcefulness were key to survival and success. Whether for military use, personal defense, or practical utility, these weapons and tools served their users by offering multiple functions in a single, often ingeniously designed package.

Today, they are valued as historical artifacts and as examples of the inventive spirit of past generations. While modern technology has largely rendered them obsolete, they remain a fascinating subject of study and a testament to the creativity of their makers.

The compilation of this small collection is not exhaustive, as the diversity and wealth of variations are too great for that, but it shows the most common combination weapon types of the 18th and 19th century.

I would like to thank the collectors Albert Detmond, Piere Dubois, Marc Leward and Walter Gross for providing photos, without which important variants of the combination weapons could not have been shown.

More books from Horst Decker

all in German language

Liebe und Krieg - Teil 1 Überlebenskunst

Liebe und Krieg sind wie Feuer und Wasser. Sie können nicht zusammen existieren, oder funktioniert das im ganz privaten Bereich doch?
Am 1. September 1939 verändert sich alles und die Möglichkeiten, das eigene Leben zu gestalten scheitert an den Niederungen des Menschseins.
ISBN 979 8848319415
Roman mit 484 Seiten - 18,00 €

Liebe und Krieg - Teil 2 Durch die Nie -

derungen des menschlichen Seins

Polen ist deutsch besetzt und die Neuordnung des dortigen Lebens beginnt, das bedeutet, aus der polnischen Bevölkerung soll ein Sklavenvolk werden, deren einzige Aufgabe es sein soll, die Lebensgrundlage für Deutschland zu verbessern.
ISBN 9798857590140
Roman mit 538 Seiten - 18,00 €

Liebe und Krieg - Teil 3 Wer jetzt nicht flieht

Die Besetzung Polens ist abgeschlossen. Ein Teil wird annektiert, der Rest erhält vorerst nur eine deutsche Regierung.
Polnische Arbeitskräfte werden als Zwangsarbeiter nach Deutschland gebracht, für Juden werden Gettos errichtet und Konzentrationslager geplant.
ISBN 979-8333913913968

Roman mit **558 Seiten - 19,50 €**

Roman zur deutschen Nachkriegszeit
"Kericht" ein Kunstwort aus Hen**ker** und **Richt**er, dem "Land der Richter und Henker", das zwischen 1945 und 1949 auf dem Gebiet der heutigen Bundesrepublik existierte. In dem ehemalige NS-Funktionäre wieder nach der Macht griffen. Die Existenz einer partiellen Rechtlosigkeit, machte vieles einfacher, nur nicht das Leben. -

ISBN 979 8848568226, Roman mit 738 Seiten, 22,- €

"Ich habe mich nur der Kunst gewidmet"

Monografie -

Ein Künstler, man sollte denken, ein emphatischer Mensch. Seine Heimatstadt ehrt ihn durch Widmung einer Straße.

Dann verraten seine Feldpostbriefe, dass er 1944 ein Arbeitslager für Juden geleitet und ca. 5000 von ihnen in den Tod geführt hat.

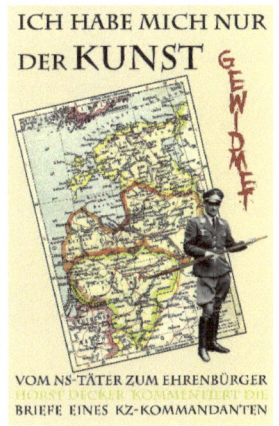

ISBN 978 3938969496, Sachbuch mit 387 Seiten - 19,95€

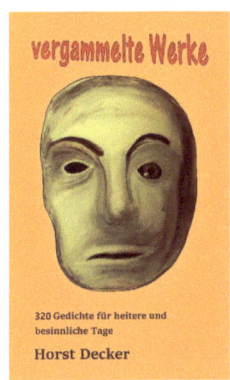

Gedichtband "Vergammelte Werke"

320 gereimte Gedichte aus allen Lebensbereichen; Humor, Politik, Liebe, Trauer etc.

Paperback: ISBN 979 832 0047 829

227 Seiten, 11,99 €

Hardcover: ISBN 979 832 017 2804

251 Seiten, 22,00 €

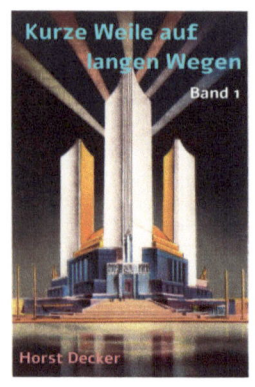

Kurze Weile auf langen Reisen

Kurzgeschichten, wahre Begebenheiten und Fiktionen
Band 1: ISBN: 979 830 3320024,
291 Seiten
Band 2: ISBN: 979 830 3548343,
293 Seiten
Band 3: ISBN: 979 830 3440722,
284 Seiten

jeweils 12,50 €

www.ingramcontent.com/pod-product-compliance
Lightning Source LLC
Chambersburg PA
CBHW040217220526
45473CB00001B/17